MEMOIRS
OF THE
AMERICAN MATHEMATICAL SOCIETY

Number 116

CONSTRUCTIVE MEASURE THEORY

by

ERRETT BISHOP and HENRY CHENG

Published by the
American Mathematical Society
Providence, Rhode Island
1972

International Standard Book Number 0-8218-1816-3
Library of Congress Catalog Number 52-42839
AMS 1970 Subject Classifications 02-E99, 28-A30

Printed in the United States of America

CONSTRUCTIVE MEASURE THEORY

by

Errett Bishop and Henry Cheng
University of California, San Diego

Received by the editors October 2, 1970.

ABSTRACT

The authors treat the Daniel integral constructively. The Daniell approach permits a general and flexibile theory that constitutes a significant advance over previous treatments of the constructive theory of measure and integration. Not only is a measure constructed from an integral, but an integral is constructed from a measure, and the two constructions are compared. Convergence theorems, Fubini's theorem, and integrals in locally compact spaces are among the topics treated.

INTRODUCTION

We hope that the present paper is close to being a definitive treatment of constructive measure theory. Historically, the first constructive theory of measure was given by Brouwer, who confined himself to the rudiments of the theory of Lebesgue measure on $\mathbb{R}^n$. Technically, Brouwer's treatment was intuitionistic, rather than constructive, because at certain points he invoked the theory of free choice sequences; however, it is not hard to constructivize his arguments. Brouwer's work was extended by various Dutch intuitionists, mostly in unpublished theses. A general and fairly complete development of constructive measure theory on locally compact spaces was given in [2]. The technical task was greatly simplified by the introduction of the notion of a complemented set. In the present paper, we give a much more natural and powerful theory, using the ideas of Daniell [4], as extended by various authors, notably M. H. Stone. The reason the Daniell approach was not used in [2] is that the classical (i.e., non-constructive) formulation of the axioms for the Daniell integral, as given for instance in [1], is not adequate to the development of a constructive theory, and the proper constructive formulation is not easy to find; in particular, it had not been found when [2] was written. Our first task is to give the proper constructive formulation of the axioms. Having done this, we develop the theory as much as possible along classical lines. Of course, the classical material must be recast at every step, and in many places, where classically it is enough to invoke the least upper bound principle, estimates are needed. As an example, the useful Lemma 4.5 would be superfluous from the classical point of view. The first major difficulty comes in proving the existence of sufficiently many integrable sets. (An integrable set is a complemented set whose characteristic function is integrable. A complemented set is an ordered pair $A \equiv (A^1, A^2)$ of sets, with $x \in A^1$ and $y \in A^2$. The characteristic function χ_A of A is that function with domain $A^1 \cup A^2$ equal to 1 on A^1 and 0 on A^2. Thus complemented sets arise naturally in the Daniell theory, whereas in [2] their introduction was perhaps somewhat mysterious.) The construction of sufficiently many integrable sets is accomplished by means of the theory of profiles, which is perhaps the most novel and interesting protion of constructive measure theory. Besides being central to our treatment of measure theory, it can be used to demonstrate a fundamental result on the existence of compact sets (Theorem 7.7), which was proved in [2] by other methods.

Having demonstrated the existence of integrable sets, we go on to study measurability and convergence in measure, and derive the monotone and dominated convergence theorems of Lebesgue. So far, we have been concerned with integration spaces, that is with linear functionals on certain spaces of functions. Next we focus attention on measures. Our definition of a measure space is a significant improvement on the presentation of [2], in particular reducing six axioms to four. It is shown that every integration space gives rise to a measure space, consisting of the integrable sets with the induced measure, and that conversely every measure space gives rise to an integration space. Necessary and sufficient conditions (called completeness conditions) are found for the measure space associated to an integration space associated to a measure space to be the original measure space, and vice versa. In Section 6, the theory is applied to give a simple construction of the product of two integration spaces, and a simple proof of Fubini's theorem, following closely the classical approach to these ideas via the Daniell integral.

As is well-known, the Daniell integral affords the most direct and powerful approach to the classical theory of measure. The same is true constructively. In Section 7, we consider the most important application, which itself contains many special cases, the theory of measure on a locally compact space. Essentially all of the material of [2] concerning (positive) measures on locally compact spaces is rederived, and some of it considerably strengthened, such as Theorem 7.11 on the existence of compact integrable sets. Theorem 7.4, which states that a continuous linear functional on $C(X)$ does indeed define an integration space, unfortunately turns out to be much harder to prove than the corresponding result (Lemma 1 of Chapter 6 of [2]), because more conditions need to be checked.

As to other applications, it is hoped that the constructive theory of stochastic processes, initiated in [3], will fit more naturally into the Daniell scheme. There are isolated developments in measure theory--the most important that comes to mind being Hausdorff measure--that do not fit the Daniell approach. It is probable that most of these can be treated constructively by the measure space approach (rather than the Daniell or integration space approach) developed in Section 5.

<u>Special conventions</u>. The symbol $\sum_n$, when no limits of summation are indicated, always denotes summation from $n = 1$ to $n = \infty$. The same convention will be used for the symbols $\bigcap$, $\bigcup$, $\bigvee$, and $\bigwedge$. The symbol ▮ denotes the conclusion of a proof, or the statement of a definition or a result stated without proof.

Some knowledge of the constructive theory of sets and metric spaces (Chapters 2, 3 and 4 of [2]) is needed to understand various statements and arguments in this paper.

INTEGRATION SPACES

We begin with the definition of an integration space, an abstract characterization of the concept of an integral. Then we define the set of integrable functions. The integrable functions are obtained by a process of completion, and they arise in applications involving limits.

Let X be a non-empty set with an equality relation $=$ and an inequality relation $\neq$, satisfying the following axioms,

$$\begin{array}{l} x = x \\ x = y \text{ implies } y = x \\ x = y \text{ and } y = z \text{ imply } x = z \\ x = y \text{ and } x \neq y \text{ imply } 0 = 1 \\ x = y \text{ and } y \neq z \text{ imply } x \neq z \\ x \neq y \text{ implies } y \neq x \\ x \neq y \text{ implies } x \neq z \text{ or } y \neq z \end{array}$$

for all x, y, z in X. Let $F(X)$ be the set of all ordered pairs $(f, D(f))$ such that $D(f)$ is a subset of X and f is a function from $D(f)$ to the set of real numbers $\mathbb{R}$, with the property that $x \neq y$ whenever $f(x) \neq f(y)$. We will usually represent such an ordered pair by its first element f, and refer to $D(f)$ as the <u>domain</u> of f. Two elements f_1 and f_2 in $F(X)$ are <u>equal</u>, written $f_1 \equiv f_2$, if $D(f_1) = D(f_2)$ and $f_1(x) = f_2(x)$ for all $x \in D(f_1)$. Functional operations can also be defined on $F(X)$. Suppose $(f, D(f))$ is in $F(X)$. Then $(|f|, D(|f|))$ is the element in $F(X)$ with $D(|f|) = D(f)$ and $(|f|)(x) = |f(x)|$ whenever $x \in D(|f|)$. Similarly we define αf, $\max\{f, \alpha\}$ and $\min\{f, \alpha\}$ for any fixed real number α. In particular $f^+ \equiv \max\{f, 0\}$ and $f^- \equiv -\min\{f, 0\}$. f is said to be non-negative if $f(x) \geq 0$ for all $x \in D(f)$. If $(g, D(g))$ is also an element of $F(X)$, then $(f, D(f)) + (g, D(g))$ is that element $(f + g, D(f + g))$ of $F(X)$ such that $D(f + g) = D(f) \cap D(g)$ and $(f + g)(x) = f(x) + g(x)$ for all $x \in D(f + g)$. Similarly fg, $\max\{f, g\}$ and $\min\{f, g\}$ are defined to be elements of $F(X)$.

Definition 1.1. A triple (X, L, I) is an integration space if X is a non-empty set with equality and inequality relations, L a subset of $F(X)$, and I a function from L to $\mathbb{R}$ such that the following four properties hold.

(1) If $f, g \in L$ and $\alpha, \beta \in \mathbb{R}$, then $\alpha f + \beta g$, $|f|$, $\min\{f, 1\}$ are in L and $I(\alpha f + \beta g) = \alpha I(f) + \beta I(g)$.

(2) If f is a function in L and $\{f_n\}_{n=1}^{\infty}$ is a sequence of non-negative functions in L such that $\sum_n I(f_n)$ converges and $\sum_n I(f_n) < I(f)$, then there exists a point $x \in X$ such that $\sum_n f_n(x)$ converges and $\sum_n f_n(x) < f(x)$. (See the Introduction for the convention governing the limits of summation.)

(3) There exists a function p in L with $I(p) = 1$.

(4) For each f in L, $\lim_{n\to\infty} I(\min\{f, n\}) = I(f)$ and $\lim_{n\to\infty} I(\min\{|f|, n^{-1}\}) = 0$. ▎

Before we go into the consequences of Properties (1) to (4), some remarks about the definition itself are appropriate here. Property (2) is the constructive analog of the continuity condition of the Daniell integral: If $\{\varphi_n\}$ is a decreasing sequence of non-negative functions in L and

$\lim_{n\to\infty} \varphi_n(x) = 0$ for all x, then $\lim_{n\to\infty} I(\varphi_n) = 0$. Indeed, we have a classical theorem that (2) implies the continuity condition. Since this fact is not needed, we shall not give a proof. Note that in Property (4), we may assume $\min\{f, n\} \in L$ for each positive integer n because $\min\{f, n\} = n \cdot \min\{n^{-1}f, 1\}$.

<u>Lemma 1.2.</u> If f_n is non-negative for each n, and $\sum_{i=1}^k I(g_i) + \sum_n I(f_n) > 0$, then $\sum_{i=1}^k g_i(x) + \sum_n f_n(x) > 0$ for some x in the common domain.

<u>Proof.</u> For all N sufficiently large

$$\sum_{n=N}^{\infty} I(f_n) < \sum_{i=1}^{k} I(g_i) + \sum_{n=1}^{N} I(f_n)$$
$$= I\left(\sum_{i=1}^{k} g_i + \sum_{n=1}^{N} f_n\right).$$

By property (2), it follows that there exists an x in the common domain with

$$\sum_{n=N}^{\infty} f_n(x) < \sum_{i=1}^{k} g_i(x) + \sum_{n=1}^{N} f_n(x). \quad \blacksquare$$

If we set $f_n \equiv p - p$ in this lemma, we get the

<u>Corollary 1.3.</u> If $\sum_{i=1}^k I(g_i) > 0$, then $\sum_{i=1}^k g_i(x) > 0$ for some x in the common domain. $\blacksquare$

This corollary is sometimes used in the weaker form, that if $\sum_{i=1}^{k} g_i(x) \geq 0$ for all x in the common domain of the $\{g_i\}$, then $\sum_{i=1}^{k} I(g_i) \geq 0$. In particular, $I(|f|) \geq |I(f)|$ for every element f in L.

<u>Corollary 1.4.</u> Let $\{f_n\}_{n=1}^{\infty}$ be a sequence of functions in L. Then $\bigcap_n D(f_n)$ is non-void.

<u>Proof.</u> $I(p) + \sum_n I(f_n - f_n) > 0$ since $I(p) = 1$ and $f_n - f_n$ is non-negative. By Lemma 1.2, there exists $x \in \bigcap_n D(f_n - f_n) = \bigcap_n D(f_n)$. ▮

Therefore the domain of any function in L is non-void.

<u>Lemma 1.5.</u> If f and g are functions in L, then $\max\{f, g\}$ and $\min\{f, g\}$ are in L.

<u>Proof.</u> Clearly $(f - g)^+ = \frac{1}{2}(f - g + |f - g|)$ is in L. It follows that $\max\{f, g\} = g + (f - g)^+$ and $\min\{f, g\} = -\max\{-f, -g\}$ are in L. ▮

Hence f^+ and f^- are in L whenever f is in L. Since $f \equiv f^+ - f^-$, we see that $I(f) = I(f^+) - I(f^-)$. The converse is also true: If f^+ and f^- are in L, then $f \equiv f^+ - f^-$ is also in L.

As it turns out, the set L is not sufficiently rich for measure-theoretic problems. We are led to enlarge L.

<u>Definition 1.6.</u> An <u>integrable function</u> consists of an ordered pair $\left(f, \{f_n\}_{n=1}^{\infty}\right)$, where $f \in F(X)$ and $\{f_n\}$ is a sequence in L, such that $\sum_n I(|f_n|)$ exists and $f(x) = \sum_n f_n(x)$ whenever $\sum_n |f_n(x)|$ converges. The <u>integral</u> of f is then defined by $I(f) = \sum_n I(f_n)$. ▮

The set of integrable functions will be denoted by L_1, with the equality relation $(f, \{f_n\}) \equiv (g, \{g_n\})$ if $f \equiv g$ as elements of $F(X)$. L_1 becomes a subset of $F(X)$ under the natural mapping $(f, \{f_n\}) \to f$. Wherever there is no confusion, we will denote an integrable function $(f, \{f_n\})$ by its first element f and call the sequence $\{f_n\}$ the <u>representation</u> of f.

Observe that I is a function on L_1, in other words, $I(f) = I(g)$ whenever $(f, \{f_n\}) \equiv (g, \{g_n\})$. This means that for any integrable function $(f, \{f_n\})$ and any sequence $\{g_n\}$ of elements of L such that $\sum_n I(|g_n|)$ converges and such that $f(x) = \sum_n g_n(x)$ whenever $\sum_n |g_n(x)|$ converges, we have

$$\sum_n I(f_n) = \sum_n I(g_n).$$

This equality is a consequence of the following

<u>Lemma 1.7.</u> Suppose $\{f_n\}$ is a sequence in L such that $\sum_n I(|f_n|)$ converges, and $\sum_n f_n(x) \geq 0$ whenever $\sum_n |f_n(x)|$ converges. Then $\sum_n I(f_n) \geq 0$.

<u>Proof.</u> Let $A \equiv \left\{x \in X : \sum_n |f_n(x)| \text{ converges}\right\}$. Then the series $\sum_n f_n^+$ and $\sum_n f_n^-$ converge on A and for each $x \in A$,

$$\sum_n f_n^+(x) - \sum_n f_n^-(x) = \sum_n f_n(x) \geq 0.$$

Let N be any positive integer. Suppose $\sum_n I(f_n^+) < \sum_{n=1}^N I(f_n^-)$. Then for some sufficiently large positive integer M we have

$$\sum_n I(f_n^+) + \sum_{n=N}^{\infty} I(f_n^-) < \sum_{n=1}^N I(f_n^-).$$

By property (2),

$$\sum_n f_n^+(x) < \sum_{n=1}^N f_n^-(x) \le \sum_n f_n^-(x)$$

for some $x \in A$. This contradiction implies that $\sum_n I(f_n^+) \ge \sum_{n=1}^N I(f_n^-)$. Since this inequality holds for any $N \ge 1$, we conclude $\sum_n I(f_n^+) \ge \sum_n I(f_n^-)$. Finally $\sum_n I(f_n) = \sum_n I(f_n^+) - \sum_n I(f_n^-) \ge 0$. ∎

L is naturally a subset of L_1 by the mapping

$$f \to (f, \{f, 0\cdot f, 0\cdot f, \cdots\}).$$

Furthermore for each $f \in L$, the value $I(f)$ is the same whether f is considered as belonging to L or L_1.

We can also define functional operations on L_1. Let $(f, \{f_n\})$ and $(g, \{g_n\})$ be integrable functions. Then $(f, \{f_n\}) + (g, \{g_n\})$ is the integrable function $(f + g, \{\varphi_n\})$, with $f + g$ the sum of f and g as elements of $F(X)$ and $\{\varphi_n\} \equiv \{f_1, g_1, f_2, g_2, \cdots\}$. For any real number α, $\alpha(f, \{f_n\})$ is defined to be $(\alpha f, \{\alpha f_n\})$. We let $|(f, \{f_n\})|$ be the integrable function $(|f|, \{\psi_n\})$, where

$$\{\psi_n\} \equiv \{|f_1|, f_1, -f_1, |f_1 + f_2| - |f_1|, f_2, -f_2, |f_1 + f_2 + f_3| - |f_1 + f_2|, \cdots\}.$$

We then have

Lemma 1.8. $I(|f|) = \lim_{n\to\infty} I\left(\left|\sum_{k=1}^{n} f_k\right|\right)$ for every integrable function $(f, \{f_k\})$. ▌

The seemingly superfluous terms $f_1, -f_1, \cdots, f_n, -f_n \cdots$ are included in the representation of $|f|$ to make sure that x is in the domain of f whenever $\sum_n |\psi_n(x)|$ converges. For the same reason, we define $\min\{(f, \{f_n\}), 1\}$ to be the integrable function $(\min\{f, 1\}, \{\lambda_n\})$, where

$$\{\lambda_n\} \equiv \{\min\{f_1, 1\}, f_1, -f_1, \min\{f_1 + f_2, 1\} - \min\{f_1, 1\}, f_2, -f_2, \cdots\}.$$

It will be seen that (X, L_1, I) is an integration space (Theorem 1.18). It is natural to ask whether a second application of Definition 1.6 leads to a still wider class of functions. The answer is no (Theorem 1.16). An explanation of this fact depends on a discussion of the domains of integrable functions.

Definition 1.9. A subset of X is called a full set if it contains a countable intersection of domains of integrable functions. ▌

Lemma 1.10. Every full set A contains the domain of some integrable function.

Proof. Let $\{f_n\}$ be a sequence of integrable functions with

$$D \equiv \bigcap_n D(f_n) \subset A.$$

Suppose each f_n has a representation $\{f_{nk}\}_{k=1}^{\infty}$ by functions belonging to L. Then let $\delta_n \equiv 1 + \sum_k I(|f_{nk}|)$ and $\{\varphi_m\}$ be the rearrangement of the double sequence $\{(2^n\delta_n)^{-1}f_{nk}\}$ into a single sequence. Then $\sum_m I(|\varphi_m|)$ converges. If we let f be the element of $F(X)$ whose domain is the domain of convergence of the series $\sum_m |\varphi_m|$, and whose value at x is $\sum_m \varphi_m(x)$, it follows that $(f, \{\varphi_m\}) \in L_1$ and $D(f) \subset D \subset A$. ▌

Proposition 1.11. Suppose f and g are integrable functions and $f(x) \leq g(x)$ for all x in a full set A ($f \leq g$ on A in concise notation). Then $I(f) \leq I(g)$.

Proof. By Lemma 1.10, A contains the domain $D(h)$ of some integrable function h. Then $f' \equiv f + h - h$ and $g' \equiv g + h - h$ are integrable functions. Furthermore $D(f') \cap D(g') = D(f') = D(g') = D(h)$ and $f'(x) \leq g'(x)$ for all x in $D(h)$ by hypothesis. Hence $I(f') \leq I(g')$ by Lemma 1.7. Since $I(f) = I(f')$ and $I(g) = I(g')$ by definition, we obtain $I(f) \leq I(g)$. ▌

In particular, if $f = g$ on A, then $I(f) = I(g)$. We can actually prove a more general statement that will be useful later.

Corollary 1.12. Let $f : D(f) \to \mathbb{R}$ be an element of $F(X)$ and g an integrable function. Suppose $f(x) = g(x)$ for all x in a full set A. Then f is also an integrable function and $I(f) = I(g)$.

Proof. A contains $D(h)$, for some h in L_1. Then $f(x) = g(x) + h(x) - h(x)$ for every point $x \in D(h)$. Now $g + h - h$ is an integrable function. Let $\{\varphi_n\}$ be its representation. Obviously, $\{\varphi_n\}$ is also a representation of f, and thus f is also integrable, with $I(f) = I(g)$. ▌

We can approach these ideas in a different way, via the notion of a norm on L_1.

Definition 1.13. The norm of an integrable function is $\|f\| \equiv I(|f|)$. ▌

Proposition 1.14. If f and g are integrable functions, then the following statements are equivalent:

(a) $f = g$ on a full set.

(b) $\|f - g\| = 0$.

Proof. Suppose $f = g$ on a full set A. Then $|f - g| = 0$ on A. Therefore $\|f - g\| = 0$.

Converesly, suppose $\|f - g\| = 0$. Let $\{\varphi_k\}$ in L be a representation of $f - g$. By Lemma 1.8, $\lim_{n\to\infty} I\left(\left|\sum_{k=1}^{n} \varphi_k\right|\right) = 0$. For each $n \in \mathbb{Z}^+$ choose an integer $k(n)$ such that

$$I(|\varphi_1 + \cdots + \varphi_{k(n)}|) < 2^{-n}$$

and $k(m) < k(n)$ whenever $m < n$. The series $\sum_n |\varphi_1 + \cdots + \varphi_{k(n)}|$ converges on a full set B_1. Write $B_2 \equiv \left\{x \in X : \sum_k |\varphi_k(x)| \text{ converges}\right\}$. Therefore $\lim_{n\to\infty} |\varphi_1(x) + \cdots + \varphi_{k(n)}(x)| = 0$ for all x in $B_1 \cap B_2$. But this means $f - g = \sum_k \varphi_k = 0$ on the full set $B_1 \cap B_2$. ▍

It is not difficult to see that the norm $\|\cdot\|$ induces a pseudo-metric ρ on L_1 by the formula $\rho(f, g) \equiv \|f - g\|$. This pseudo-metric becomes a metric if we define a new notion of equality on L_1 by taking $f = g$ to mean that $\|f - g\| = 0$, or equivalently that $f = g$ on some full set. This equality, called <u>equality almost everywhere</u>, is more important than the equality $\equiv$ used heretofore. Hence, for any two elements f and g in $F(X)$, we define $f = g[f \le g]$ to mean $f = g[f \le g]$ on some full set.

Now we are ready to state and prove the "completeness theorem" for integrable functions. The method of proof revolves around a simple lemma.

<u>Lemma 1.15.</u> For each $\varepsilon > 0$, each f in L_1 has a representation $\{\varphi_n\}$ with $\sum_n I(|\varphi_n|) < I(|f|) + \varepsilon$.

<u>Proof.</u> Let $\{f_n\}$ be any representation of f. Choose an integer $N \ge 1$ with $\sum_{n=N+1}^{\infty} I(|f_n|) < \frac{1}{2}\varepsilon$. Write $\{\varphi_n\} \equiv \left\{\sum_{n=1}^{N} f_n, f_{N+1}, f_{N+2}, \cdots\right\}$. Then for each $m > N$ we have

$$\sum_n I(|\varphi_n|) \le I\left(\left|\sum_{n=1}^{N} f_n\right|\right) + \frac{1}{2}\varepsilon$$

$$= I\left(\left|\sum_{n=1}^{m} f_n\right|\right) + I\left(\left|\sum_{n=1}^{N} f_n\right| - \left|\sum_{n=1}^{m} f_n\right|\right) + \frac{1}{2}\varepsilon$$

$$\leq I\left(\left|\sum_{n=1}^{m} f_n\right|\right) + I\left(\left|\sum_{n=N+1}^{m} f_n\right|\right) + \frac{1}{2}\varepsilon$$

$$\leq I\left(\left|\sum_{n=1}^{m} f_n\right|\right) + \sum_{n=N+1}^{m} I(|f_n|) + \frac{1}{2}\varepsilon$$

$$< I\left(\left|\sum_{n=1}^{m} f_n\right|\right) + \varepsilon .$$

Since $I\left(\left|\sum_{n=1}^{m} f_n\right|\right) \to I(|f|)$ as $m \to \infty$, the lemma follows. ▌

Theorem 1.16. Suppose $\{f_n\}$ is a sequence of integrable functions and $\sum_n I(|f_n|)$ converges. Then there exists a full set A and an integrable function f such that $\sum_n |f_n(x)|$ exists for every $x \in A$ and $f = \sum_n f_n$ on A. For each such f, we have $\lim_{N\to\infty} \left\|f - \sum_{n=1}^{N} f_n\right\| = 0$.

Proof. By Lemma 1.15, we can find for each $n \geq 1$ a representation $\{f_{nk}\}_{k=1}^{\infty}$ of the integrable function f_n such that $\sum_k I(|f_{nk}|) < I(|f_n|) + 2^{-n}$. Let $A \equiv \left\{x \in X : \sum_n \sum_k |f_{nk}(x)| \text{ converges}\right\}$. Clearly A is a full set. Then the function f defined by the conditions

(a) $D(f) \equiv A$

(b) $f(x) \equiv \sum_n \sum_k f_{nk}(x)$ for all $x \in A$

is an integrable function, and any enumeration of the double sequence $\{f_{nk}\}_{k,n=1}^{\infty}$ gives a representation of f. Moreover $\sum_n |f_n|$ converges on A

and so $f = \sum_n f_n$ on A.

Now consider any such f, and any $N \in \mathbb{Z}^+$. Then $|f - \sum_{n=1}^{N} f_n| \le \sum_{n=N+1}^{\infty} |f_n|$ on the full set A. Let h be an integrable function with $D(h) \subset A$, and let $\{h_m\}$ represent $h - h$. Let $\{g_m\}$ represent $|f - \sum_{n=1}^{N} f_n|$. Then

$$\sum_{n=N+1}^{\infty} \sum_k |f_{nk}(x)| - \sum_m g_m(x) + \sum_m h_m(x)$$

$$\ge \sum_{n=N+1}^{\infty} |f_n(x)| - \left| f(x) - \sum_{n=1}^{N} f_n(x) \right| \ge 0$$

at any point x at which the series $\sum_{n=N+1}^{\infty} \sum_k |f_{nk}(x)|$, $\sum_m |g_m(x)|$ and $\sum_m |h_m(x)|$ all converge. Then Lemma 1.7 gives

$$\left\| f - \sum_{n=1}^{N} f_n \right\| = \sum_m I(g_m)$$

$$\le \sum_{n=N+1}^{\infty} \sum_k I(|f_{nk}|)$$

$$\le \sum_{n=N+1}^{\infty} \left(I(|f_n|) + 2^{-n} \right).$$

The last sum approaches 0 as $N \to \infty$. ▌

If $\{g_n\}$ is any representation of an integrable function g (so that each $g_n \in L$), it follows from Theorem 1.16 that $\lim_{N\to\infty} \left\|g - \sum_{n=1}^{N} g_n\right\| = 0$. Thus we get the very useful

<u>Corollary 1.17.</u> L is dense in L_1. ▌

Although $L \subset L_1$, the latter set of functions retains all the analytic properties of the former. This can be stated formally as a theorem.

<u>Theorem 1.18.</u> (X, L_1, I) is an integration space.

<u>Proof.</u> We have to show that Properties (1) through (4) of Definition 1.1 are valid for the set L_1 of integrable functions. Property (1) is naturally valid for L_1. To verify (2), suppose h is a non-negative integrable function and $\{h_n\}_{n=1}^{\infty}$ is a sequence of non-negative integrable functions such that $\sum_n I(h_n) < I(h)$. Then $\sum_n I(h_n) + 3\varepsilon < I(h)$ for some $\varepsilon > 0$. By Lemma 1.15, we can choose for each $n \geq 1$ a sequence $\{\varphi_{nk}\}_{k=1}^{\infty}$ in L that represents h_n, with $\sum_k I(|\varphi_{nk}|) < I(h_n) + \varepsilon 2^n$. Aslo choose a sequence $\{\psi_k\}_{k=1}^{\infty}$ in L to represent h and a positive integer K such that $I(h) < I\left(\left|\sum_{k=1}^{K} \psi_k\right|\right) + \varepsilon$ and $\sum_{k=K+1}^{\infty} I(|\psi_k|) < \varepsilon$. Then

$$\sum_n \left(\sum_k I(|\varphi_{nk}|)\right) + \sum_{k=K+1}^{\infty} I(|\psi_k|) < I\left(\left|\sum_{k=1}^{K} \psi_k\right|\right).$$

Property (2), for the original space L, gives

$$\sum_n \left(\sum_k |\varphi_{nk}(x)| \right) + \sum_{k=K+1}^{\infty} |\psi_k(x)| < \left| \sum_{k=1}^{K} \psi_k(x) \right|$$

for some point $x \in X$. Hence

$$\sum_n h_n(x) < \left| \sum_{k=1}^{K} \psi_k(x) \right| - \left| \sum_{k=K+1}^{\infty} \psi_k(x) \right|$$

$$\leq \left| \sum_k \psi_k(x) \right| = h(x).$$

Thus Property (2) is valid for the space L_1.

Since $L \subset L_1$, the function p in L such that $I(p) = 1$ serves the same purpose in L_1. Thus Property (3) is valid for L_1.

Finally we shall use Corollary 1.17 to prove that L_1 has Property (4). Let f be any element of L_1. Since $\{I(\min\{f, n\})\}_{n=1}^{\infty}$ is a monotone non-decreasing sequence bounded above by $I(f)$, to show it converges to $I(f)$ it is enough to show that for each $\varepsilon > 0$ there exists $N \geq 1$ with $I(\min\{f, N\}) \geq I(f) - 3\varepsilon$. Choose $g \in L$ with $\|f - g\| \leq \varepsilon$. Then

$$|I(f) - I(g)| = |I(f - g)| \leq I(|f - g|) = \|f - g\| \leq \varepsilon.$$

Also, for each $n \geq 1$,

$$|I(\min\{f, n\}) - I(\min\{g, n\})|$$

$$\leq I(|\min\{f, n\} - \min\{g, n\}|) \leq I(|f - g|) \leq \varepsilon.$$

Since $I(\min\{g, n\}) \to I(g)$ as $n \to \infty$, there exists $N \geq 1$ with $I(\min\{g, N\}) \geq I(g) - \varepsilon$. Hence $I(\min\{f, N\}) \geq I(\min\{g, N\}) - \varepsilon \geq I(g) - 2\varepsilon \geq I(f) - 3\varepsilon$. To show that $I(\min\{|f|, n^{-1}\}) \to 0$ as $n \to \infty$, for each $\varepsilon > 0$ we must find an integer N with $I(\min\{|f|, N^{-1}\}) \leq 3\varepsilon$. This is done by choosing g in L with $\||f| - |g|\| \leq \varepsilon$, and then choosing the integer N with $I(\min\{|g|, N^{-1}\}) \leq \varepsilon$. ▌

INTEGRABLE SETS

As an application of the theory of integration spaces we will show that the integrable functions induce a measure on a certain family of complemented subsets in X. Then the behavior of the measure with respect to the basic set operations will be investigated.

Definition 2.1. Let (X, L, I) be an integration space. A complemented set in X consists of an ordered pair $A \equiv (A^1, A^2)$ of subsets A^1 and A^2 of X such that $x \neq y$ whenever $x \in A^1$ and $y \in A^2$. We say $x \in A$ if $x \in A^1$ and $x \in -A$ if $x \in A^2$. Complemented sets (A^1, A^2) and (B^1, B^2) are equal if $A^1 = B^1$ and $A^2 = B^2$. ▌

If A is a complemented set in X, then A^1 $[A^2]$ will always denote the first [second] component of A. Next we define operations on complemented sets.

Definition 2.2. Let $A \equiv (A^1, A^2)$ and $B \equiv (B^1, B^2)$ be complemented sets.

(a) $A \wedge B \equiv \left(A^1 \cap B^1, (A^1 \cap B^2) \cup (A^2 \cap B^1) \cup (A^2 \cap B^2)\right)$.

(b) $A \vee B \equiv \left((A^1 \cap B^1) \cup (A^1 \cap B^2) \cup (A^2 \cap B^1), A^2 \cap B^2\right)$.

(c) A is a subset of B, written $A < B$, if $A^1 \subset B^1$ and $A^2 \supset B^2$.

(d) $-B \equiv (B^2, B^1)$ is the complement of B.

(e) $A - B \equiv A \wedge (-B)$.

(f) Let $\{(C_k^1, C_k^2)\}_{k=1}^{\infty}$ be a sequence of complemented sets. Then

$$(C^1, C^2) \equiv \bigwedge_k (C_k^1, C_k^2)$$

is the complemented set defined by

$$C^1 \equiv \bigcap_k C_k^1$$

and

$$C^2 \equiv \left[\bigcap_k (C_k^1 \cup C_k^2)\right] \cap \left[\bigcup_k C_k^2\right].$$

$$(D^1, D^2) \equiv \bigvee_k (C_k^1, C_k^2)$$

is the complemented set defined by

$$D^1 \equiv \left[\bigcap_k (C_k^1 \cup C_k^2)\right] \cap \left[\bigcup_k C_k^1\right]$$

and

$$D^2 \equiv \bigcap_k C_k^2. \quad \blacksquare$$

Intuitively, $A \wedge B$ and $A \vee B$ represent the intersection and union of A and B. They are used instead of the simplier definitions

$$A \cap B \equiv (A^1 \cap B^1, A^2 \cup B^2) \quad \text{and} \quad A \cup B \equiv (A^1 \cup B^1, A^2 \cap B^2)$$

given in [2], because they correspond directly to operations on functions. This fact is revealed by the following definition.

Definition 2.3. The characteristic function χ_A of a complemented set $A \equiv (A^1, A^2)$ is the function defined on the subset $A^1 \cup A^2$ of X with

$$\chi_A(x) = \begin{cases} 1 & \text{whenever } x \in A \\ 0 & \text{whenever } x \in -A. \end{cases}$$

A complemented set A is integrable if its characteristic function χ_A is an integrable function, and we define the measure of A to be

$$\mu(A) \equiv I(\chi_A). \quad \blacksquare$$

It is easy to verify that

$$\chi_{A \wedge B} \equiv \min \{\chi_A, \chi_B\} \equiv \chi_A \cdot \chi_B$$

and

$$\chi_{A \vee B} \equiv \chi_A + \chi_B - \chi_{A \wedge B}$$

as elements of $F(X)$. These equalities will expedite the proofs of the elementary properties of integrable sets.

Proposition 2.4. Let A and B be integrable sets. Then $A \vee B$ and $A \wedge B$ are integrable sets and

$$\mu(A) + \mu(B) = \mu(A \vee B) + \mu(A \wedge B).$$

Proof. Let χ_A and χ_B be the characteristic functions of A and B respectively. Then

$$\chi_{A\wedge B} \equiv \min\{\chi_A, \chi_B\}$$

and

$$\chi_{A\vee B} \equiv \chi_A + \chi_B - \chi_{A\wedge B}$$

are both integrable functions. Moreover

$$\chi_A + \chi_B \equiv \chi_A + \chi_B - \chi_{A\wedge B} + \chi_{A\wedge B}$$

$$\equiv \chi_{A\vee B} + \chi_{A\wedge B}. \quad ▌$$

Proposition 2.5. Let C and D be complemented sets. If C and $C \wedge D$ are integrable, then $C - D$ is integrable and $\mu(C) = \mu(C \wedge D) + \mu(C - D)$.

Proof. $\chi_{C-D} \equiv \chi_C - \chi_{C\wedge D}$ implies $C - D$ is an integrable set and $\mu(C - D) = \mu(C) - \mu(C \wedge D)$. ▌

The relationships between full sets and integrable sets are worth examining, although the results hold little surprise.

Proposition 2.6. Let F be a full set and $A \equiv (A^1, A^2)$ an integrable set with $\mu(A) > 0$. Then $A^1 \cap F$ has at least one element.

Proof. Let F contain the domain $D(g)$ of some integrable function g. Consider the integrable function

$$h \equiv I(\chi_A)(I(|g|) + 1)^{-1} |g|.$$

Then $I(h) < I(\chi_A)$ and so there exists $x \in X$ such that

$$0 \leq I(\chi_A)(I(|g|) + 1)^{-1} |g(x)| < \chi_A(x).$$

Hence $\chi_A(x) = 1$, and therefore $x \in A^1 \cap F$. ▌

Proposition 2.7. If $A \equiv (A^1, A^2)$ is an integrable set and $\mu(A) = 0$, then A^2 is a full set.

Proof. Since $\mu(A) = I(\chi_A) = 0$, $I(n\chi_A) = nI(\chi_A) = 0$ for all $n \in \mathbb{Z}^+$. Therefore $\sum_n I(|n\chi_A|)$ is convergent and so $\sum_n n\chi_A$ defines an integrable function f, whose domain $D(f)$ is the full set $\left\{x \in X : \sum_n n\chi_A(x) \text{ converges}\right\}$. Let $x \in D(f)$. Then $x \in D(\chi_A)$, and the assumption $\chi_A(x) \neq 0$ implies $0 = 1$. Hence $\chi_A(x) = 0$. This means $x \in A^2$. Thus A^2 is a full set because $D(f) \subset A^2$. ▌

Proposition 2.7 is the constructive substitute of the classical statement that a full set is the complement of a null set. In the case of an integrable set whose measure we don't know to be equal to zero or not, we still have the following result.

Proposition 2.8. $A^1 \cup A^2$ is a full set for any integrable set (A^1, A^2).

Proof. The domain of any integrable function is a full set. ▮

Corollary 2.9. If $E < F$ are integrable sets, then $\mu(E) \leq \mu(F)$.

Proof. Since $\mu(A) \geq 0$ for every integrable set A, we have $\mu(E \wedge F) \leq \mu(F)$ by Proposition 2.5. But $\chi_E = \chi_{E\wedge F}$ on the full set $(E^1 \cup E^2) \cap (F^1 \cup F^2)$. By Proposition 1.11, $\mu(E) = \mu(E \wedge F)$. ▮

Proposition 2.10. Let $\{A_k\}_{k=1}^{\infty}$ be a sequence of integrable sets. Then

(a) if $\alpha \equiv \lim_{n\to\infty} \mu\left(\bigvee_{k=1}^{n} A_k\right)$ exists, then $\bigvee_k A_k$ is integrable with measure α,

(b) if $\sum_k \mu(A_k)$ converges, then $\bigvee_k A_k$ is integrable and $\mu\left(\bigvee_k A_k\right) \leq \sum_k \mu(A_k)$,

(c) if $\beta \equiv \lim_{n\to\infty} \mu\left(\bigwedge_{k=1}^{n} A_k\right)$ exists, then $\bigwedge_k A_k$ is integrable with measure β.

Proof. (a) Suppose α exists. Let $B_n \equiv \bigvee_{k=1}^{n} A_k$. We define $\varphi_0 \equiv 0$ and

$$\varphi_n \equiv \varphi_{n-1} + \chi_n - \min\{\varphi_{n-1}, \chi_n\}$$

for each $n \geq 1$, where χ_n is the characteristic function of A_n. Then φ_n

is the characteristic function of B_n, and hence $I(\varphi_n) \to \alpha$ as $n \to \infty$. Now let $\lambda_n \equiv \varphi_n - \varphi_{n-1}$ for $n \geq 1$. Then $0 \leq \lambda_n \leq 1$,

$$\sum_n I(|\lambda_n|) = \sum_n I(\lambda_n) = \lim_{n\to\infty} I(\varphi_n) = \alpha,$$

and so $\sum_n \lambda_n$ is an integrable function f, with $I(f) = \alpha$. In fact, f is the characteristic function of $\bigvee_k A_k$.

(b) Suppose $\sum_k \mu(A_k)$ converges. For arbitrary integers $P, Q \geq 1$, we obtain from Proposition 2.4 the following inequalities:

$$0 \leq \mu\left(\bigvee_{k=1}^{P+Q} A_k\right) - \mu\left(\bigvee_{k=1}^{P} A_k\right) \leq \mu\left(\bigvee_{k=P+1}^{P+Q} A_k\right) \leq \sum_{k=P+1}^{P+Q} \mu(A_k).$$

Hence $\lim_{n\to\infty} \mu\left(\bigvee_{k=1}^{n} A_k\right)$ exists, and (a) implies that $\bigvee_k A_k$ is integrable. Since $\mu\left(\bigvee_{k=1}^{n} A_k\right) \leq \sum_{k=1}^{n} \mu(A_k)$ for all $n \in \mathbf{Z}^+$, we conclude that $\mu\left(\bigvee_k A_k\right) \leq \sum_k \mu(A_n)$.

(c) The argument is analogous to the one given for (a) and is therefore omitted. ▎

THE THEORY OF PROFILES

In the next section we prove the famous monotone convergence and dominated convergence theorems of Lebesgue. It will be crucial to our arguments to know the existence of a sufficient number of integrable sets of the form $\{x \in X : f(x) \geq \alpha\}$ for any given integrable function f. This section is devoted entirely to this existence problem.

Definition 3.1. Let $[a, b]$ be a compact proper $(a < b)$ interval on $\mathbb{R}$. Let $C[a, b]$ denote the set of continuous functions on $[a, b]$. Let $F \subset C[a, b]$ be any subset with the properties:

(a) For any f in F, $0 \leq f \leq 1$.

(b) The constant functions 0 and 1 are in F.

(c) If $a \leq u < v \leq b$, then there exists f in F with $f(t) = 0$ for $t \leq u$ and $f(t) = 1$ for $t \geq v$.

Finally, let $\lambda : F \to \mathbb{R}$ be any non-decreasing function, that is, $\lambda(f) \leq \lambda(g)$ whenever $f(x) \leq g(x)$ for all $x \in [a, b]$. Then (F, λ) is called a profile over $[a, b]$. ▌

The obvious example of a profile is obtained by letting $\lambda(f)$ equal the ordinary Riemann integral $\int_a^b f(x)\, dx$ of a continuous function $f : [a, b] \to R$. It will be helpful to keep this example in mind.

Definition 3.2. Let $[u, v]$ be a proper subinterval of $[a, b]$ and (F, λ) be a profile over $[a, b]$. For any fixed $\delta > 0$, the expression

$$P\big([u, v]\big) < \delta$$

means that there exist $f^1, f^2 \in F$ with

(a) $f^1(t) = 0$ for $t \le v$,

(b) $f^2(t) = 1$ for $t \ge u$,

(c) $\lambda(f^2) - \lambda(f^1) < \delta$.

The expression $P'\big([u, v]\big) < \delta$ means that there exists $\alpha > 0$ such that

$$P\big([\max\{a, u - \alpha\}, \min\{b, v + \alpha\}]\big) < \delta. \quad \blacksquare$$

Intuitively, $P\big([u, v]\big) < \delta$ means that the profile λ varies by less than δ on $[u, v]$, and $P'\big([u, v]\big) < \delta$ means that λ varies by less than δ on some slightly larger interval.

We first show that the value of P on most small intervals is small.

Lemma 3.3. If (F, λ) is a profile over $[a, b]$, $\varepsilon > 0$, and $n \ge 1$, with $(n + 1)\varepsilon > \lambda(1) - \lambda(0)$, then for each $\delta > 0$, there exist an integer N and points $a = a_0 < a_1 < \dots < a_N = b$ and non-negative integers $M_1, \dots, M_N$ such that

(a) $a_i - a_{i-1} \le \delta \quad (1 \le i \le N)$,

(b) $\sum_{i=1}^{N} M_i = n$,

(c) $P'\big([a_{i-1}, a_i]\big) < \big(M_i + 1\big)\varepsilon \quad (1 \le i \le N)$.

Proof. Choose an integer K and points $a = b_0 < \dots < b_K = b$ such that $(b_i - b_{i-1}) < \frac{1}{2}(n + 1)^{-1}\delta$ for $1 \le i \le K$. Let

$$\sigma \equiv \frac{1}{4} \min \big\{b_i - b_{i-1} : 1 \le i \le K\big\}.$$

Let $f_0 \equiv 1$, $f_{K+1} \equiv 0$, and for $1 \le i \le K$ choose $f_i \in F$ such that $f_i(t) = 0$ if $t \le b_{i-1} + \sigma$ and $f_i(t) = 1$ if $t \ge b_i - \sigma$. Set $\alpha_i \equiv \lambda(f_{i-1}) - \lambda(f_{i+1})$, for $1 \le i \le K$. Partition the integers i $(1 \le i \le K)$ into two sets A and B, with $\alpha_i < \varepsilon$ for all $i \in A$ and $\alpha_i > (n+1)^{-1}(\lambda(1) - \lambda(0))$ for all $i \in B$. Let the subsequence $a = a_0 < a_1 < \cdots < a_N = b$ of $b_0, \ldots, b_K$ consist of all b_i, such that either $i = 0$ or $i = K$ or $i \in A$ or $i + 1 \in A$. Consider $j \in \{1, \ldots, N\}$. Then $a_j = b_i$ for some value of i. Either (i) $i \in A$ or (ii) $i \in B$. In case (i), $a_{j-1} = b_{i-1}$ by definition of the subsequence $a_0, \ldots, a_N$. Also, $P'\big([a_{j-1}, a_j]\big) < \varepsilon$ because $\alpha_i = \lambda(f_{i-1}) - \lambda(f_{i+1}) < \varepsilon$. In case (i) we therefore define $M_j \equiv 0$, thereby satisfying (c). Next consider case (ii). Then $i \in B$, and $a_{j-1} = b_k$ for some k with $k < i$. We claim that $a_j - a_{j-1} = b_i - b_k \le \delta$. Suppose the contrary: $b_i - b_k > \delta$. By the definition of the sequence $b_0, \ldots, b_K$, it follows that $i - k > 2(n+1)$. Consider any integer m with $k < m < i$. If $m \in A$, then b_m belongs to the sequence $a_0, \ldots, a_N$, and in addition $a_{j-1} = b_k < b_m < b_i = a_j$. This contradiction proves that $m \in B$. Also $i \in B$, since otherwise b_{i-1} would belong to the sequence $a_0, \ldots, a_N$. Thus $m \in B$ for $k < m \le i$. Hence

$$\sum \{\alpha_m : k < m \le i\} > 2(n+1)((n+1)^{-1}(\lambda(1) - \lambda(0)))$$

$$= 2(\lambda(1) - \lambda(0))$$

But

$$\sum_{i=1}^{K} \alpha_i = \sum_{i=1}^{K} \lambda(f_{i-1}) - \lambda(f_{i+1}) \le 2(\lambda(1) - \lambda(0)).$$

This contradiction implies that $a_j - a_{j-1} \le \delta$, for each j in case (ii). For each such j write $\beta_j \equiv \lambda(f_k) - \lambda(f_{i+1})$. Then $P'\big([a_{j-1}, a_j]\big) \le \beta_j$. Consider a second index $j' \in \{1, \ldots, N\}$ also in case (ii), with $a_{j'-1} = b_{k'}$ and $a_{j'} = b_{i'}$, such that $j < j'$. Then $i + 1 \le k'$. Therefore $f_k \ge f_{i+1} \ge f_{k'} \ge f_{i'+1}$. It follows that $\sum' \beta_j \le \lambda(1) - \lambda(0)$, where $\sum'$ denotes summation over the set of j satisfying (ii). For each such j, choose a non-negative integer M_j' with

$$M_j'\varepsilon - N^{-1}\{(n+1)\varepsilon - (\lambda(1) - \lambda(0))\} < \beta_j < (M_j' + 1)\varepsilon$$

Then

$$\sum{}' M_j'\varepsilon < \sum{}' \beta_j + (n+1)\varepsilon - (\lambda(1) - \lambda(0)) \le (n+1)\varepsilon.$$

Therefore $\sum' M_j' \le n$. Choose $M_j \ge M_j'$ such that $\sum' M_j = n$. Since $M_j' = 0$ for all j in case (i), it follows that $\sum_{j=1}^{N} M_j = n$. Thus (a), (b) and (c) are satisfied. ▌

The intuitive meaning of the next lemma is that $t_1, \ldots, t_n$ are the only points where the profile can jump by more than ε.

Lemma 3.4. Let (F, λ) be a profile over $[a, b]$. Consider $n \geq 1$ and $\varepsilon > 0$, with $(n + 1)\varepsilon > \lambda(1) - \lambda(0)$. Then there exist points $t_1, \ldots, t_n$ in $[a, b]$, such that for each $\beta > 0$ there exists $\gamma > 0$ such that if $t \in [a, b]$ and $|t - t_i| > \beta$ for all $i \in \{1, \ldots, n\}$, then

$$P\Big([\max \{a, t - \gamma\}, \min \{b, t + \gamma\}]\Big) < \varepsilon.$$

Proof. By induction on the positive integer k, we construct for each k a finite family $\mathcal{F}_k$ of non-overlapping proper compact subintervals of $[a, b]$, a function $\varphi_k : \mathcal{F}_k \to \mathbb{Z}^+$, and a constant $\sigma_k > 0$ such that

(a) there exists a function $\gamma_k : \mathcal{F}_{k+1} \to \mathcal{F}_k$ with $I \subset \gamma_k(I)$ for each I in $\mathcal{F}_k$ and $\sum\{\varphi_{k+1}(I) : \gamma_k(I) = J\} = \varphi_k(J)$ for each $J \in \mathcal{F}_k$,

(b) $P'(I) < (\varphi_k(I) + 1)\varepsilon$ for each $I \in \mathcal{F}_k$,

(c) if $x \in [a, b] - \bigcup \mathcal{F}_k$, then $P\Big([\max \{a, x - \sigma_k\}, \min \{b, x + \sigma_k\}]\Big) < \varepsilon$,

(d) $\sum \{\varphi_k(I) : I \in \mathcal{F}_k\} = n$, and

(e) each element of $\mathcal{F}_k$ has length $\leq k^{-1}(\max \{1, b - a\})$.

We begin the construction by taking $\mathcal{F}_1 \equiv \{[a, b]\}$, $\varphi_1([a, b]) \equiv n$, and $\sigma_1 \equiv 1$. Now assume we have carried the construction through to the integer k. Consider $[a', b'] \in \mathcal{F}_k$. Since $P'\Big([a', b']\Big) < \Big(\varphi_k([a', b']) + 1\Big)\varepsilon$, there exists $f^1, f^2 \in F$ such that $f^1(t) = 0$ for $t \leq \min \{b, b' + \sigma\}$ and $f^2(t) = 1$ for $t \geq \max \{a, a' - \sigma\}$ for some $\sigma > 0$, and $\lambda(f^2) - \lambda(f^1) < \Big(\varphi_k([a', b']) + 1\Big)\varepsilon$. Let G be the subset of all functions f in F such that either $f = f^1$, or $f = f^2$, or else $f(t) = 0$ on $[a, a']$ and $f(t) = 1$ on $[b', b]$. For each f in G, let f' be the restriction of f to $[a', b']$, and define $\lambda'(f') \equiv \lambda(f)$. These functions form a family F' such that (F', λ') is a profile over $[a', b']$ and

$$\lambda'(1) - \lambda'(0) = \lambda(f^2) - \lambda(f^1) < \left(\varphi_k([a', b']) + 1\right)\varepsilon.$$

By Lemma 3.3, there exist points $a' = a_0 < \dots < a_N = b'$ and non-negative integers $M_1, \dots, M_n$ such that $\sum_{i=1}^N M_i = \varphi_k\left([a', b']\right)$, $a_i - a_{i-1} \leq (k+1)^{-1} \max\{1, b - a\}$, and $P'\left([a_{i-1}, a_i]\right) \subset \left(M_i + 1\right)\varepsilon$. Let $\mathcal{F}_{k+1}\left([a', b']\right)$ consist of all intervals $[a_{i-1}, a_i]$ with $M_i > 0$. Let

$$\mathcal{F}_{k+1} \equiv \bigcup\{\mathcal{F}_{k+1}(I) : I \in \mathcal{F}_k\}, \quad \gamma_k\left([a_{i-1}, a_i]\right) = [a', b'],$$

and $\varphi_{k+1}\left([a_{i-1}, a_i]\right) = M_i$.

Conditions (a), (b), (d) and (e) are satisfied by $\mathcal{F}_{k+1}$, φ_{k+1} and γ_k. It remains to define σ_{k+1} and check (c). To each interval $I \equiv [a_{i-1}, a_i]$ for which $M_i = 0$ there exists $\omega(I) > 0$ with

$$(*) \qquad P[\max\{a, a_{i-1} - \omega(I)\}, \min\{b, a_i + \omega(I)\}] < \varepsilon$$

Let ω be the inf of all such $\omega(I)$. Let σ_{k+1} be the minimum of $\frac{1}{2}\omega$ and $\frac{1}{2}\sigma_k$. Consider $x \in [a, b] - \bigcup\mathcal{F}_{k+1}$. Then there exists y with

$|x - y| < \sigma_{k+1}$ with either

(i) $y \in \bigcup \mathcal{F}_k - \bigcup \mathcal{F}_{k+1}$ or

(ii) $y \in [a, b] - \bigcup \mathcal{F}_k$.

In case (i), $y \in I \equiv [a_{i-1}, a_i] \subset [a', b'] \in \mathcal{F}_k$, with $M_i = 0$. Since $x - \sigma_{k+1} \geq y - 2\sigma_{k+1} \geq a_{i-1} - \omega(I)$ and $x + \sigma_{k+1} \leq a_i + \omega(I)$, inequality (c) follows from (*). In case (ii), inequality (c) follows from

$$P[\max \{a, y - \sigma_k\}, \min \{b, y + \sigma_k\}] < \varepsilon.$$

The induction is complete.

Now, again by induction on k, choose $I_1^k, \ldots, I_n^k \in \mathcal{F}_k$ such that (i) $I_j^{k+1} \subset I_j^k$ for $1 \leq j \leq n$, such that (ii) for $1 \leq j \leq n - 1$ either $I_j^k = I_{j+1}^k$ or else I_j^k and I_{j+1}^k are non-overlapping with I_j^k to the left, and such that (iii) each element I of $\mathcal{F}_k$ occurs $\varphi_k(I)$ times in the sequence $I_1^k, \ldots I_n^k$. Since I_j^k is non-void and $\operatorname{diam}\left(I_j^k\right) \to 0$ as $k \to \infty$, $\bigcap_k I_j^k$ consists of a single point t_j, with $a \leq t_1 \leq t_2 \leq \ldots \leq t_n \leq b$.

Finally, let $\beta > 0$ be given. Choose $k \in \mathbf{Z}^+$ such that

$$k^{-1}\left(\max \{1, b - a\}\right) < \beta.$$

Let $t \in [a, b]$ and $|t - t_j| \geq \beta$ for all $j \in \{1, \ldots, n\}$. Then $t \in [a, b] - \bigcup \mathcal{F}_k$. Therefore $P\left([\max \{a, t - \sigma_k\}, \min \{b, t + \sigma_k\}]\right) < \varepsilon$, so we may take $\gamma = \sigma_k$. ▌

Theorem 3.5. If (F, λ) is a profile over $[a, b]$, then all except countably many points t in $[a, b]$ are smooth, in the sense that there exists a number $\bar{\lambda}(t)$ such that for each $\varepsilon > 0$, there exists $\delta > 0$ such that $|\lambda(f) - \bar{\lambda}(t)| < \varepsilon$ whenever $f(x) = 1$ for all $x \geq t + \delta$ and $f(x) = 0$ for all $x \leq t - \delta$.

Proof. Construct an increasing sequence $\{n(k)\}_{k=1}^{\infty}$ of positive integers such that $(n(k) + 1)k^{-1} > \lambda(1) - \lambda(0)$ and $\left\{t_1^k, \ldots, t_{n(k)}^k\right\}$ are the excluded points in the preceding lemma. Let T be the collection $\left\{t_j^k : 1 \leq j \leq n(k), 1 \leq k < \infty\right\}$. Then t is smooth if t is unequal to each of the points T. To prove this, we must exhibit the number $\bar{\lambda}(t)$. To this end, for each k choose a positive constant δ_k such that $|t - t_j^k| > \delta_k$ whenever $1 \leq j \leq n(k)$. Thus there exists a positive number γ_k with

$$P\Big([\max \{a, t - \gamma_k\}, \min \{b, t + \gamma_k\}]\Big) < k^{-1}.$$

Furthermore we may take $\{\gamma_k\}$ to be a strictly decreasing sequence converging to 0. As a result, there exist sequences $\left\{f_k^1\right\}_{k=1}^{\infty}$ and $\left\{f_k^2\right\}_{k=1}^{\infty}$ of functions in F such that

(a) $f_k^1(t) = 0$ for $t \leq \min \{b, t + \gamma_k\}$, and $f_k^1 \leq f_{k+1}^1$,

(b) $f_k^2(t) = 1$ for $t \geq \max \{a, t - \gamma_k\}$, and $f_k^2 \geq f_{k+1}^2$,

(c) $0 \leq \lambda(f_k^2) - \lambda(f_k^1) \leq k^{-1}$.

Therefore $\{\lambda(f_k^1)\}$ and $\{\lambda(f_k^2)\}$ are both Cauchy sequences and share a common limit $\overline{\lambda}(t)$.

Let $\varepsilon > 0$ be given. Choose k so large that $k^{-1} < \varepsilon$. Let $\delta \equiv \gamma_k$. Suppose $f : [a, b] \to \mathbb{R}$ is a continuous function such that $f(x) = 1$ for all $x \geq t + \delta$ and $f(x) = 0$ for all $x \leq t - \delta$. Then $f_k^1 \leq f \leq f_k^2$. Since $\lambda(f_k^1) \leq \overline{\lambda}(t) \leq \lambda(f_k^2)$ and $0 \leq \lambda(f_k^2) - \lambda(f_k^1) \leq k^{-1}$, we have $|\lambda(f) - \overline{\lambda}(t)| \leq k^{-1} < \varepsilon$. ▍

We are now ready to apply the theory to show the existence of many integrable sets.

<u>Theorem 3.6.</u> Let (X, L, I) be an integration space and h an integrable function. Then for all except countably many $t > 0$, the sets

$$A \equiv \big(\{x : h(x) \geq t\}, \{x : h(x) < t\}\big)$$

and

$$B \equiv \big(\{x : h(x) > t\}, \{x : h(x) \leq t\}\big)$$

are integrable and have the same measure.

<u>Proof.</u> Since $\mathbb{R}^+ = \bigcup_n (n^{-1}, n)$, it is sufficient to show that for all except countably many points t in a given interval (a, b), with $0 < a < b$, A and B are integrable. To this end, for every pair of real numbers u, v such that $0 < u < v$, let

$$f_{u,v}(x) = \begin{cases} 0 & \text{if } x \le u \\ \frac{x-u}{v-u} & \text{if } u \le x \le v \\ 1 & \text{if } x \ge v \end{cases}$$

and $\lambda_0(f_{u,v}) = I(f_{u,v} \circ h)$. (Note that $f_{u,v} \circ h$ is an integrable function because $f_{u,v} \circ h \equiv \min \{(v - u)^{-1} (h - \min \{h, u\}), 1\}$.) Choose real numbers $\alpha, \beta, \gamma, \delta$ such that $0 < \alpha < \beta < a < b < \gamma < \delta$. Let F_0 be the set of all f such that $f = f_{\alpha,\beta}$ or $f = f_{\gamma,\delta}$ or $f = f_{u,v}$ with $a \le u < v \le b$. Let F be the set of f' such that f' is the restriction to $[a, b]$ of some function f in F_0. Set $\lambda(f') \equiv \lambda_0(f)$. Then (F, λ) is a profile over $[a, b]$. By Theorem 3.5, all except countably many t in (a, b) are smooth.

Let t be smooth. Then $\lim_{\varepsilon\to 0} I(f_{t-\varepsilon,t} \circ h) = \overline{\lambda}(t)$. Choose a positive integer m with $m^{-1} < t - a$, and for each $n \ge m$ write $t(n) \equiv t - n^{-1}$. Then by Theorem 1.16 the series $\sum_{n=m}^{\infty} \varphi_n$ defined by $\varphi_m \equiv f_{t(m),t} \circ h$ and $\varphi_n \equiv f_{t(n),t} \circ h - f_{t(n-1),t} \circ h$ for $n > m$ converges on the full set $D(f) \equiv \left\{x : \sum_{n=m}^{\infty} |\varphi_n(x)| \text{ converges}\right\}$ to an integrable function f such that $I(f) = \overline{\lambda}(t)$.

We claim that $f(x) \equiv \chi_A(x)$ for all x in $D(f)$. Consider any $x \in D(f)$. Then $x \in D(h)$ and either $f(x) < 1$ or $f(x) > 0$. If $f(x) < 1$, then $f_{t(n),t} \circ h(x) < 1$ for some n. Hence $h(x) < t$ and so $f(x) = 0$. If $f(x) > 0$, then $h(x) \ge t - n^{-1}$ for all $n \ge m$. Hence $h(x) \ge t$ and so $f(x) = 1$. By Corollary 1.12, χ_A is integrable and $I(\chi_A) = \overline{\lambda}(t)$. The proof that $I(\chi_B) = \overline{\lambda}(t)$ is essentially the same. ▌

MEASURABLE FUNCTIONS AND THE CONVERGENCE OF INTEGRALS

We are now prepared to study the concept of convergence in the class of integrable functions. Our study begins with the introduction of the class of measurable functions. The relationship between measurable and integrable functions is examined and a sufficient condition for measurability is derived. This condition leads naturally to a definition of convergence.

Definition 4.1. Let (X, L, I) be an integration space. A function $h : D(h) \to \mathbb{R}$ defined on a subset $D(h)$ of X is *measurable* if

$$\mathrm{mid}\,(-n, \chi_A h, n) \equiv \max\,\{\min\,\{\chi_A h, n\}, -n\}$$

is an integrable function for every integrable set A and every $n \geq 1$. ▌

If $A \equiv (A^1, A^2)$ is an integrable set, then we consider $(A^1 \cup A^2) \cap D(h)$ to be the domain of $\chi_A h$. It follows from the above definition that $D(h)$ is a full set.

The measurable functions share many of the properties of integrable functions. For instance, $\min\,\{f, g\}$, $\max\,\{f, g\}$ and $|f|$ are measurable whenever f and g are measurable. Also f is measurable if and only if f^+ and f^- are measurable. The following proposition shows that every integrable function is measurable.

Proposition 4.2. Let A be an integrable set and f an integrable function. Then $\chi_A f$ is an integrable function.

Proof. Since $\lim_{n\to\infty} I(\min\,\{f, n\}) = I(f)$, there exists a sequence of increasing positive integers $n_1 < n_2 < \ldots$ such that

$$I(f) - I(\min\{f, n_k\}) \le 2^{-k}.$$

Setting $n_0 = 0$, we obtain the representation

$$\chi_A f = \sum_k \min\left\{(n_k - n_{k-1})\chi_A,\ f - \min\{f, n_{k-1}\}\right\}.$$

Hence $\chi_A f$ is integrable. ▌

Of course, not every measurable function is integrable. However the lemma below suggests a condition under which it is.

Lemma 4.3. Let f be a non-negative integrable function. Then

$$c \equiv \sup\left\{I(\min\{\chi_A f, n\}) : A \text{ is an integrable set and } n \ge 1\right\}$$

and

$$c' \equiv \sup\left\{I(\chi_A f) : A \text{ is an integrable set}\right\}$$

exist, and $c = c' = I(f)$.

Proof. From Property (4), it is clear that c exists if and only if c' exists, in which case they are equal. Thus we need only prove that c' exists, and $c' = I(f)$. Clearly the set $T \equiv \{I(\chi_A f) : A \text{ is an integrable set}\}$ is bounded by $I(f)$. Therefore we seek to find a sequence from T that converges to $I(f)$. By Theorem 3.6, we know that there is a sequence $\{\alpha_n\}$ of real numbers such that $0 < \alpha_{n+1} < \alpha_n$, $\lim_{n\to\infty} \alpha_n = 0$ and

$$A_n \equiv \left(\{x : f(x) \geq \alpha_n\}, \{x : f(x) < \alpha_n\}\right)$$

is an integrable set for each n. We have $\lim_{n\to\infty} I(f - \min\{f, \alpha_n\}) = I(f)$, since $I(f) = I(\min\{f, \alpha_n\}) + I(f - \min\{f, \alpha_n\})$ and $\lim_{n\to\infty} I(\min\{f, \alpha_n\}) = 0$. But $f - \min\{f, \alpha_n\} \leq \chi_n f \leq f$ on the full set $A_n^1 \cup A_n^2$, where χ_n is the characteristic function of A_n. Hence

$$I\left(f - \min\{f, \alpha_n\}\right) \leq I\left(\chi_n f\right) \leq I(f)$$

and we conclude that

$$\lim_{n\to\infty} I\left(\chi_n f\right) = I(f). \quad \blacksquare$$

<u>Proposition 4.4.</u> A non-negative measurable function f is integrable if the integration space is $\sum$-finite, meaning that there exists a sequence $B_1 < B_2 < \cdots$ of integrable sets such that $\bigcup_n B_n^1$ is a full set, and if the constant c of Lemma 4.3 exists.

<u>Proof.</u> Let $\{A_n\}$ be a sequence of integrable sets such that $A_1 < A_2 < \cdots$ and $0 \leq c - I(\min\{\chi_n' f, n\}) \leq 2^{-n}$ where χ_n' is the characteristic function of A_n. Write $C_n \equiv A_n \vee B_n$. For each n, let χ_n be the characteristic function of C_n. Define the functions $\lambda_1 \equiv \min\{\chi_1 f, 1\}$, and $\lambda_n \equiv \min\{\chi_n f, n\} - \min\{\chi_{n-1} f, n-1\}$ for $n \geq 2$. Then $g \equiv \sum_n \lambda_n$ is an

integrable function, that equals f on the full set

$$D(f) \cap \left(\bigcup_n C_n^1\right) \cap \bigcap_n \left(C_n^1 \cap C_n^2\right).$$

By Corollary 1.12, f is integrable and $I(f) = I(g) = c$. ▌

For the next theorem and corollary, we need a simple lemma, whose proof is left to the reader.

Lemma 4.5. Let φ and ψ be maps from the same set S into the real numbers, such that for arbitrary s_1 and s_2 in S there exists s_3 in S with

$$0 \leq \varphi(s_3) - \varphi(s_i) \leq \psi(s_3) - \psi(s_i)$$

for $i = 1$ and $i = 2$. Then if $\{\psi(s) : s \in S\}$ has a least upper bound, so does $\{\varphi(s) : s \in S\}$. ▌

Theorem 4.6. A measurable function f on a $\sum$-finite integration space is integrable if

$$c \equiv \sup \left\{ I(\min\{\chi_A|f|, n\}) : A \text{ is an integrable set and } n \geq 1 \right\}$$

exists.

Proof. By proposition 4.4, it is enough to show that

$$c' \equiv \sup \left\{ I(\min\{\chi_A f^+, n\}) : (A, n) \in S \right\}$$

and

$$c'' \equiv \sup \left\{ I(\min \{\chi_A f^-, n\}) : (A, n) \in S \right\}$$

both exist, where S is the cartesian product of the class of integrable sets and the positive integers. For each $s \equiv (A, n)$ in S set $\varphi(s) \equiv I(\min \{\chi_A f^+, n\})$ and $\psi(s) \equiv I(\min \{\chi_A |f|, n\})$. Consider $s_1 \equiv (A_1, n_1)$ and $s_2 \equiv (A_2, n_2)$ in S. Set $s_3 \equiv (A_1 \vee A_2, n_1 + n_2)$. Then

$$\begin{aligned} 0 \le \varphi(s_3) - \varphi(s_i) &= \varphi(s_3) - \varphi((A_1 \vee A_2, n_i)) + \varphi((A_1 \vee A_2, n_i)) - \varphi(s_i) \\ &\le \psi(s_3) - \psi((A_1 \vee A_2, n_i)) + \psi((A_1 \vee A_2, n_i)) - \psi(s_i) \\ &= \psi(s_3) - \psi(s_i) \end{aligned}$$

for $i = 1$ and $i = 2$. Then c' exists by Lemma 4.5. A similar proof shows that c'' exists. ▌

Corollary 4.7. If the integration space is $\sum$-finite, if f is measurable, and if $|f| \le g$ for some integrable g, then f is integrable.

Proof. It is enough to consider the case $f \ge 0$, and to show that the constant c of Lemma 4.3 exists.

By Lemma 4.3,

$$c_0 \equiv \sup \left\{ I(\min \{\chi_A g, n\}) : A \text{ integrable}, n \ge 1 \right\}$$

exists. Write $S \equiv \{(A, n) : A \text{ integrable}, n \geq 1\}$, $\varphi((A, n)) \equiv I(\min\{\chi_A f, n\})$, and $\psi((A, n)) \equiv I(\min\{\chi_A g, n\})$. If $s_i \equiv (A, m)$ and $s_2 \equiv (B, n)$ are arbitrary elements of S, we take $s_3 \equiv (A \vee B, m + n)$. Then it is easily shown that, for $i = 1$ and $i = 2$,

$$0 \leq \varphi(s_3) - \varphi(s_i) \leq \psi(s_3) - \psi(s_i).$$

Since c_0 exists, it follows that c exists. ▌

Measurable sets are defined analogously to integrable sets.

Definition 4.8. A complemented set S is measurable if its characteristic function χ_S is measurable. ▌

Thus S is measurable if $\chi_S \cdot \chi_A \equiv \chi_{S \wedge A}$ is integrable for each integrable set A. The measurable sets are an algebra (i.e., they are closed under the finite set operations). If A is a measurable set and f an integrable function, then $\chi_A f$ is integrable, and we write $I_A(f) \equiv I(\chi_A f)$. In case the integration space is $\sum$-finite, the proof consists in taking $f \geq 0$ and using Lemma 4.5 to show that

$$\sup\left\{I(\min\{\chi_B \chi_A f, n\}) : B \text{ integrable}, n \geq 1\right\}$$

exists because $\sup\{I(\min\{\chi_B f, n\}) : B \text{ integrable}, n \geq 1\}$ exists. With a little extra work, the $\sum$-finiteness hypothesis can be removed.

Proposition 4.9. Let $h : D(h) \to \mathbb{R}$ be defined on a full set $D(h) \subset X$. For each integrable set A, each $n \geq 1$, and each $\varepsilon > 0$, let there exist a measurable function f and an integrable set $B < A$ such that $\mu(A - B) < \varepsilon$ and $\text{mid}(-n, h, n) = f$ on B. Then h is measurable.

Proof. Under the hypothesis, f may be considered integrable because $\text{mid}(-n, h, n) = f$ on B implies $\text{mid}(-n, h, n) = \text{mid}(-n, \chi_A f, n)$ on B, and the last function is integrable. Suppose that the integer $n \geq 1$ and the integrable set A are fixed. Choose a sequence $\{B_k\}$ of integrable sets and a sequence $\{f_k\}$ of integrable functions such that $B_1 < A$, $B_2 < A - B_1, \ldots$, $\mu(A - B_1 - \ldots - B_k) \leq 2^{-k}$, and $\text{mid}(-n, h, n) = f_k$ on B_k. Let χ_k be the characteristic function of B_k. Then $\sum_k \chi_k f_k$ is an integrable function f since $I(|\chi_k f_k|) \leq n\, \mu(B_k) \leq n2^{-k+1}$. Let $A_1 = A - B_1$ and in general $A_k = A - B_1 - \ldots - B_k$. Then $\bigwedge_k A_k$ is integrable and $\mu\left(\bigwedge_k A_k\right) = 0$. Hence $\left\{x : x \in -\bigwedge_k A_k = \bigvee_k - A_k\right\}$ is a full set. Since $-A_k = -A \vee B_1 \vee \ldots \vee B_k$, $\text{mid}(-n, \chi_A h, n) = \chi_A f$ on this full set. Therefore h is measurable. ∎

Theorem 4.10. Let $h : D(h) \to \mathbb{R}$ be a function defined on a full set $D(h) \subset X$. Suppose for each integrable set A, each $n \geq 1$, and each $\varepsilon > 0$ that there exists a measurable function f and an integrable set $B < A$ such that $\mu(A - B) < \varepsilon$ and $|\text{mid}(-n, h, n) - f| < \varepsilon$ on B. Then h is measurable.

Proof. Again, we may assume f is integrable. Choose a sequence $\{B_k\}$ of integrable sets such that $B_k < A$, $\mu(A - B_k) < \varepsilon\, 2^{-k}$, and $|\text{mid}(-n, h, n) - f_k| < \varepsilon\, 2^{-k}$ on B_k for some integrable function f_k. Observe that, for $\ell < m$,

$$0 \le \mu\left(\bigwedge_{k=1}^{\ell} B_k\right) - \mu\left(\bigwedge_{k=1}^{m} B_k\right)$$

$$= \left[\mu(A) - \mu\left(A - \bigwedge_{k=1}^{\ell} B_k\right)\right] - \left[\mu(A) - \mu\left(A - \bigwedge_{k=1}^{m} B_k\right)\right]$$

$$= \mu\left(A - \bigwedge_{k=1}^{m} B_k\right) - \mu\left(A - \bigwedge_{k=1}^{\ell} B_k\right)$$

$$\le \mu\left(A - \bigwedge_{k=\ell+1}^{m} B_k\right)$$

$$\le \sum_{k=\ell+1}^{m} \mu(A - B_k) \le \varepsilon\, 2^{-\ell}.$$

Therefore $B \equiv \bigwedge_k B_k$ is integrable and $\mu(A - B) < \varepsilon$. Furthermore $\|\chi_B f_m - \chi_B f_\ell\| < \varepsilon\, 2^{-\ell+1}\, \mu(B)$, and hence $\chi_B f_k$ converges to an integrable function f on some full set S. Let T be the complemented set (S, ϕ). Then mid $(-n, h, n) = f$ on $T \wedge B$. Since $\mu(T \wedge B) = \mu(B)$, we use Proposition 4.9 to finish the proof. ▌

<u>Definition 4.11.</u> Let $\{f_n\}_{n=1}^{\infty}$ be a sequence of measurable functions and $f : D(f) \to \mathbb{R}$ a function which is defined on a full subset $D(f)$ of X. The sequence $\{f_n\}$ <u>converges to</u> f <u>in measure</u> if for each integrable set A and each $\varepsilon > 0$ there exists a positive integer N such that for each $n \ge N$ there exists an integrable set B with $B^1 \subset A^1 \cap D(f) \cap D(f_n)$, $\mu(A - B) < \varepsilon$, and $|f(x) - f_n(x)| < \varepsilon$ for all $x \in B$. ▌

The limit function f is obviously measurable, by Theorem 4.10. We will show that this limit function is essentially unique. Two measurable functions f and g are <u>equal</u> if

$$\text{mid}\,(-n, \chi_A f, n) = \text{mid}\,(-n, \chi_A g, n)$$

as integrable functions for all integrable sets $\{A\}$ and positive integers $\{n\}$.

<u>Proposition 4.12.</u> If $\{f_n\}$ converges in measure to both f and g, then $f = g$ as measurable functions.

<u>Proof.</u> Consider any integrable set A and any $\varepsilon > 0$. Choose $N \geq 1$ so that there exist integrable sets B and C such that

(a) $B^1 \subset A^1 \cap D(f) \cap D(f_N)$,

(b) $\mu(A - B) < \frac{1}{2}\varepsilon$,

(c) $|f(x) - f_N(x)| < \frac{1}{2}\varepsilon$ for all $x \in B$,

(d) $C^1 \subset A^1 \cap D(g) \cap D(f_N)$,

(e) $\mu(A - C) < \frac{1}{2}\varepsilon$, and

(f) $|g(x) - f_N(x)| < \frac{1}{2}\varepsilon$ for all $x \in C$.

Since $\varepsilon > \mu(A - B) + \mu(A - C) \geq \mu(A - B \wedge C)$, we let $E \equiv B \wedge C$ and see that $|f(x) - g(x)| < \varepsilon$ for all $x \in E$. This implies that if n is any positive integer,

$$|\text{mid}\,(-n, \chi_A f, n) - \text{mid}\,(-n, \chi_A g, n)| < \varepsilon$$

on the integrable set E. Hence

$$I\left(|\text{mid }(-n, \chi_A f, n) - \text{mid }(-n, \chi_A g, n)|\right)$$

$$\leq \varepsilon\, \mu(E) + n\, \mu(A - E) < (\mu(E) + n)\varepsilon.$$

Since ε can be made arbitrarily small, we conclude that

$$\text{mid }(-n, \chi_A f, n) = \text{mid }(-n, \chi_A g, n)$$

as integrable functions. ∎

The convergence of functions is intimately related to the convergence of their integrals. A rigorous statement of one such relationship is the <u>monotone convergence theorem</u>.

<u>Theorem 4.13.</u> Let $f_1 \leq f_2 \leq \ldots$ be a monotone sequence of integrable functions. If $c \equiv \lim_{n\to\infty} I(f_n)$ exists, then $\{f_n\}$ converges in measure to an integrable function f, and $I(f) = c$. Conversely, if $\{f_n\}$ converges in measure to an integrable function f, then $\lim_{n\to\infty} I(f_n) = I(f)$.

<u>Proof.</u> Suppose $\lim I(f_n) = c$. Define $\lambda_1 = f_1$ and $\lambda_k = f_k - f_{k-1}$ for $k \geq 2$. Then the sum $\sum_k \lambda_k$ is an integrable function f with $I(f) = c$. Clearly $f_n \leq f$ for all $n \geq 1$. To show that f is the limit of $\{f_n\}$ in measure, let A be an integrable set and $\varepsilon > 0$. Choose $N \geq 1$ such that $0 \leq c - I(f_n) < \frac{1}{2}\varepsilon^2$ for all $n \geq N$. Choose a number δ in the interval

$(\frac{1}{2}\varepsilon, \varepsilon)$ so that $S \equiv \left(\{x : f(x) - f_n(x) \geq \delta\}, \{x : f(x) - f_n(x) < \delta\}\right)$ is an integrable set. If we set $B \equiv A - S$, then

$$\mu(A - B) = I(\chi_{A-B}) \leq \delta^{-1} I(f - f_n)$$

$$= \delta^{-1}(c - I(f_n)) < \varepsilon.$$

Also $0 \leq f(x) - f_n(x) < \delta < \varepsilon$ for each $x \in B$. Therefore $\{f_n\}$ converges to f in measure.

Now suppose $\{f_n\}$ converges to f in measure. First we show that $I(f)$ is an upper bound of the sequence $\{I(f_n)\}$. Choose a strictly decreasing sequence $\{\alpha_k\}_{k=1}^{\infty}$ of positive real numbers converging to 0 such that, for each k,

$$A_k \equiv \left(\{x : f_n(x) - f(x) \geq \alpha_k\}, \{x : f_n(x) - f(x) < \alpha_k\}\right)$$

is an integrable set, where n is any fixed positive integer. For each pair (k, j) of positive integers we can find an integer $p > n$ and an integrable set $B < A_k$ such that $\mu(A_k - B) < \alpha_k 2^{-j}$ and $|f(x) - f_p(x)| < \alpha_k 2^{-j}$ for all $x \in B$. Since $f_p \geq f_n$, the set B must be void. Hence

$$\mu(A_k) = \mu(A_k \wedge B) + \mu(A_k - B) < \alpha_k 2^{-j}.$$

It follows that $\mu(A_k) = 0$ for all $k \in \mathbf{Z}^+$. Hence $f \geq f_n$ on the full set

$$\{x : x \in -A_k \text{ for all } k \geq 1\}$$

and so $I(f) \geq I(f_n)$.

To show that $I(f)$ is indeed the supremum of $\{I(f_n)\}$, let $\varepsilon > 0$ be given. By Lemma 4.3, there exists an integrable set A such that $I_{-A}(f) < \varepsilon$. Choose $n \geq 1$ such that $I(f) - I(\min \{f, n\}) < \frac{1}{2}\varepsilon$. Then $I_C(f) \leq \frac{1}{2}\varepsilon + I_C(\min \{f, n\}) \leq \frac{1}{2}\varepsilon + n\mu(C) < \varepsilon$ for every integrable set C with $\mu(C) < \delta \equiv (2n)^{-1}\varepsilon$. Also there exist $N \geq 1$ and an integrable set $B < A$ such that $\mu(A - B) < \delta$ and $|f(x) - f_N(x)| < \varepsilon(\mu(A) + 1)^{-1}$ on B. Hence

$$\begin{aligned} 0 \leq I(f) - I(f_N) &\leq I_{-A}(f) + I_A(f) - I_B(f_N) \\ &= I_{-A}(f) + I_{A-B}(f) + I_B(f - f_N) < 3\varepsilon. \end{aligned}$$

Therefore $\lim_{n\to\infty} I(f_n) = I(f)$. ▌

When the convergent sequence of functions is not monotone, some other condition is needed to guarantee the convergence of integrals. Lebesgue, of course, produced such a condition in his dominated convergence theorem. The proof of this theorem will depend on a simple lemma.

<u>Lemma 4.14.</u> Let $\{f_n\}_{n=1}^{\infty}$ be a sequence of non-negative integrable functions converging in measure to 0. For each $\varepsilon > 0$ let there exist an integrable set A, an integer $N \geq 1$, and $\delta > 0$ such that $I_B(f_n) < \varepsilon$ whenever $n \geq N$ and B is a measurable set with $\mu(A \wedge B) < \delta$. Then $\lim_{n\to\infty} I(f_n) = 0$.

Proof. For each $\varepsilon > 0$, choose A, N, δ as described. In addition, choose N so large that for each $n \geq N$ there exists an integrable set $C < A$ such that $\mu(A - C) < \delta$ and $|f_n| < \varepsilon(\mu(A) + 1)^{-1}$ on C. Let $B \equiv -C$, which is measurable. Then $\mu(A \wedge B) = \mu(A - C) < \delta$. By the hypothesis, $I_B(f_n) < \varepsilon$. Also $I_C(f_n) < \varepsilon(\mu(A) + 1)^{-1} \mu(C) < \varepsilon$. Hence $I(f_n) = I_C(f_n) + I_{-C}(f_n) < 2\varepsilon$. ∎

Theorem 4.15. Let the sequence $\{f_n\}_{n=1}^{\infty}$ of integrable functions converge in measure to the integrable function f, and suppose there exists an integrable function g such that $|f_n| \leq g$ for all $n \geq 1$. Then $\lim_{n\to\infty} I(f_n) = I(f)$.

Proof. To show that $\lim_{n\to\infty} I(f_n) = I(f)$, it is enough to show that $\lim_{n\to\infty} I(|f_n - f|) = 0$. Let $\varepsilon > 0$ be given. As in the proof of Theorem 4.13, choose an integrable set A with $I_{-A}(g) < 4^{-1}\varepsilon$ and choose $\delta > 0$ such that $I_C(g) < 4^{-1}\varepsilon$ whenever C is an integrable set with $\mu(C) < \delta$. Now if B is any measurable set with $\mu(A \wedge B) < \delta$, then

$$\chi_B = \chi_{A\wedge B} + \chi_{-A\wedge B} \leq \chi_{A\wedge B} + \chi_{-A}$$

on a full set. This implies that

$$\begin{aligned} I_B(|f_n - f|) &\leq I_{A\wedge B}(|f_n - f|) + I_{-A}(|f_n - f|) \\ &\leq I_{A\wedge B}(2g) + I_{-A}(2g) < \varepsilon. \end{aligned}$$

By Lemma 4.14, $I(|f_n - f|) \to 0$ as $n \to \infty$. ∎

So far we have limited ourselves to the study of convergence in measure. Related concepts such as almost-uniform convergence and almost-everywhere convergence can also be introduced and their relationships with convergence in measure fruitfully studied. However such an investigation would not differ significantly from the work already done in [2]. Therefore we refer the reader to the relevant theorems in Chapter 7 of [2].

MEASURE FOR MEASURE

In Section 2, we defined the measures of certain complemented sets as the integrals of their characteristic functions. In this section we reverse this process. Starting with the basic properties of the set function $A \to \mu(A)$, we construct a space of integrable functions.

<u>Definition 5.1.</u> Let X be a set with equality and inequality relations. A <u>measure space</u> $(X, \mathfrak{M}, \mu)$ is a set $\mathfrak{M}$ of complemented subsets of X and a function $\mu : \mathfrak{M} \to \mathbb{R}^{0+}$ having the following properties.

(a) If A and B are in $\mathfrak{M}$, then $A \vee B$ and $A \wedge B$ are in $\mathfrak{M}$ and

$\mu(A) + \mu(B) = \mu(A \vee B) + \mu(A \wedge B)$.

(b) If A and $A \wedge B$ are in $\mathfrak{M}$, then $A - B$ is in $\mathfrak{M}$ and

$\mu(A) = \mu(A \wedge B) + \mu(A - B)$.

(c) There exists $M \in \mathfrak{M}$ with $\mu(M) > 0$.

(d) For each sequence $\{M_n\}$ of sets in $\mathfrak{M}$ such that $\lim_{n\to\infty} \mu(M_1 \wedge \ldots \wedge M_n)$ exists and is positive, the set $\bigwedge_n M_n$ is non-void. ▌

It is a simple matter to check that the integrable sets derived from an integration space form a measure space. Hence we call μ the <u>measure</u>, and $\mathfrak{M}$ the set of <u>integrable sets</u>.

Proposition 5.2 gives some very elementary facts concerning a measure space, which will be useful later.

<u>Proposition 5.2.</u> Let $(X, \mathfrak{M}, \mu)$ be a measure space.

(i) If $A \in \mathfrak{M}$ and $A^1 = \phi$, then $\mu(A) = 0$.

(ii) If $A_1, \ldots, A_n$ are in $\mathfrak{M}$, then $(\phi, F) \in \mathfrak{M}$, where $F \equiv \bigcap_{k=1}^{n} (A_k^1 \cup A_k^2)$. Moreover if $B \in \mathfrak{M}$, then $(B^1 \cap F, B^2 \cap F) \in \mathfrak{M}$ and $\mu((B^1 \cap F, B^2 \cap F)) = \mu(B)$.

<u>Proof.</u> (i) Suppose $\mu(A) \neq 0$. Let $M_n \equiv A$ for every $n \geq 1$. By (d) of Definition 5.1, A^1 is non-void. This contradiction implies $\mu(A) = 0$.

(ii) Since $A_k \in \mathfrak{M}$, $A_k - A_k \in \mathfrak{M}$ by (b) of Definition 5.1. But $A_k - A_k = (\phi, A_k^1 \cup A_k^2)$. Hence $(\phi, F) = (\phi, \bigcap_{k=1}^{n} (A_k^1 \cup A_k^2)) = \bigwedge_{k=1}^{n} (A_k - A_k) \in \mathfrak{M}$.

Let $N \equiv (\phi, F)$. Then

$$\begin{aligned}\mu(B) &= \mu(B \wedge N) + \mu(B - N)\\ &= \mu((\phi, (B^1 \cup B^2) \cap F)) + \mu((B^1 \cap F, B^2 \cap F))\\ &= \mu((B^1 \cap F, B^2 \cap F)). \quad \blacksquare\end{aligned}$$

Our goal now is to examine how a measure space gives rise to an integration space.

<u>Definition 5.3.</u> Let $(X, \mathfrak{M}, \mu)$ be a measure space. An element χ of $F(X)$ is a <u>simple function</u> if $\chi \equiv \sum_{k=1}^{n} c_k \chi_k$, where each $c_k \in \mathbb{R}$ and $\chi_1, \ldots, \chi_n$ are the characteristic functions of integrable sets $A_1, \ldots, A_n$. Since each χ_k is regarded as an element of $F(X)$, the domain $D(\chi)$ of χ is equal to $\bigcap_{k=1}^{n} (A_k^1 \cup A_k^2)$. The expression $c_1\chi_1 + \ldots + c_n\chi_n$ is called a <u>representation</u> of the simple function χ. The <u>integral</u> of χ is defined by $I(\chi) \equiv c_1\mu(A_1) + \ldots + c_n\mu(A_n)$. $\blacksquare$

We have to show that I is a function. The method of proof consists of using a combinatorial lemma on integrable sets. (See [2], page 188 for the proof.)

Lemma 5.4. Let $A(1), \ldots, A(n)$ be integrable sets and $a_1, \ldots, a_n$ be real numbers. Let P consist of all free subsets S of $\{1, \ldots, n\}$. (The subset S is free means that it is possible to decide whether or not a given integer belongs to S.) For each S in P let $A(S)$ be the integrable set

$$A(S) \equiv \left(\bigwedge_{i\in S} A(i)\right) \wedge \left(\bigwedge_{i\in -S} -A(i)\right).$$

Then the sets $\{A(S) : S \in P\}$ are disjoint (in the sense that $A(S) < -A(T)$ whenever $S \neq T$),

$$\mu(A(i)) = \sum_{i\in S} \mu(A(S)) \qquad (1 \leq i \leq n)$$

and

$$\sum_{i=1}^{n} a_i \chi_{A(i)} \equiv \sum_{S\in P}\left(\sum_{i\in S} a_i\right) \chi_{A(S)}. \quad \blacksquare$$

Theorem 5.5. The integral $I(\chi)$ of a simple function χ does not depend on the representation of χ.

Proof. Suppose $f \equiv \sum_{i=1}^{n} a_i \chi_i$ and $g \equiv \sum_{j=1}^{m} b_j \chi_j'$ are simple functions with χ_i the characteristic function of A_i and χ_j' the characteristic function of

B_j, such that $f = g$ on $D(f) \cap D(g)$. By introducing coefficients equal to 0 it suffices to consider the case $m = n$ and $B_i = A_i (1 \le i \le n)$. Define the sets $\{A(S)\}$ as in Lemma 5.4. As a result,

$$I(f) = \sum_{i=1}^{n} a_i \mu(A_i)$$

$$= \sum_{i=1}^{n} a_i \left(\sum_{i \in S} \mu(A(S)) \right)$$

$$= \sum_{S \in P} \left(\sum_{i \in S} a_i \right) \mu(A(S))$$

and similarly,

$$I(g) = \sum_{S \in P} \left(\sum_{I \in S} b_i \right) \mu(A(S)).$$

If $I(f) \ne I(g)$, then

$$\left(\sum_{i \in S} a_i - \sum_{i \in S} b_i \right) \mu(A(S)) \ne 0$$

for some $S \in P$. This implies there exists a point x in $A(S)$ such that

$$f(x) - g(x) = \left(\sum_{i \in S} a_i - \sum_{i \in S} b_i \right) \chi_{A(S)}(x) \ne 0.$$

This contradiction implies that $I(f) = I(g)$. ▮

We need two estimates on the integral for the proof of the theorem that a measure space induces an integration space.

Lemma 5.6. If f is a non-negative simple function and ε is a positive number, then there exists A in $\mathfrak{M}$ such that $A^1 \cup A^2 \subset D(f)$, $f < \varepsilon$ on A^2, and $\mu(A) \le 2\varepsilon^{-1} I(f)$.

Proof. Suppose $f \equiv \sum_{i=1}^n a_i\chi_i$, where $\{\chi_i\}$ are the characteristic functions of disjoint sets $\{A_i\}$. Partition $\{1, \ldots, n\}$ into disjoint sets S and T, with $a_i < \varepsilon$ for all $i \in S$ and $a_i > \frac{1}{2}\varepsilon$ for all $i \in T$. Define

$$B \equiv \bigvee_{i\in T} A_i \quad \text{and} \quad A \equiv (B^1 \cap D(f), B^2 \cap D(f)).$$

Then we have, by Proposition 5.2,

$$\mu(A) = \mu\left(\bigvee_{i\in T} A_i\right) \le \sum_{i\in T} \mu(A_i)$$

$$\le \sum_{i\in T} 2a_i\varepsilon^{-1} \mu(A_i) \le 2\varepsilon^{-1} I(f).$$

Now suppose $x \in -A$. Then $x \in \bigwedge_{i\in T} - A_i$. Hence

$$f(x) = \sum_{i\in S} a_i\chi_i(x) + \sum_{i\in T} a_i\chi_i(x)$$

$$= \sum_{i\in S} a_i\chi_i(x) \le \max\{a_i : i \in S\} < \varepsilon. \quad ▮$$

Lemma 5.7. Let f be a simple function. Let c be a positive upper bound for f. Let A be an integrable set, with $f(x) \leq 0$ for $x \in D(f) \cap A^2$. Then for each positive ε less than c, there exists $B \in \mathfrak{M}$ with $\mu(B) \geq c^{-1}(I(f) - 2\varepsilon\,\mu(A))$, with $B^1 \cup B^2 \subset D(f)$, and with $f > \varepsilon$ on B.

Proof. Let $f \equiv \sum_{i=1}^{n} a_i \chi_i$, where $\{\chi_i\}$ are the characteristic functions of disjoint sets $\{A_i\}$. Partition $\{1, \ldots, n\}$ into disjoint sets S and T, with $a_i > \varepsilon$ for $i \in S$ and $a_i < 2\varepsilon$ for $i \in T$. Let $E \equiv \bigvee_{i \in S} A_i$ and $B \equiv (E^1 \cap D(f), E^2 \cap D(f))$. Then $f > \varepsilon$ on B and we have the estimate

$$I(f) \leq I(\chi_A f) = \sum_{i \in S} a_i\, \mu(A_i \wedge A) + \sum_{i \in T} a_i\, \mu(A_i \wedge A)$$

$$\leq c\,\mu(B) + 2\varepsilon\,\mu(A).$$

Hence $\mu(B) \geq c^{-1}(I(f) - 2\varepsilon\,\mu(A))$. ▌

Theorem 5.8. Let $(X, \mathfrak{M}, \mu)$ be a measure space. Let L be the class of simple functions $\sum a_i \chi_i$, where $\{\chi_i\}$ are the characteristic functions of integrable sets $\{A_i\}$. Let the function $I : L \to \mathbb{R}$ be defined by

$$I\left(\sum a_i \chi_i\right) \equiv \sum a_i\, \mu(A_i).$$

Then (X, L, I) is an integration space.

Proof. We have to show that the four basic properties (1), (2), (3) and (4) of Definition 1.1 are satisfied by (X, L, I). To verify (1), let $f, g \in L$ and $\alpha, \beta \in \mathbb{R}$. By Lemma 5.4, there exist disjoint integrable sets $\{A_i\}$ such that $f = \sum_{i=1}^{m} a_i \chi_i$ and $g = \sum_{i=1}^{m} b_i \chi_i$ on $D(f) \cap D(g)$. It follows that $\alpha f + \beta g$, $|f|$ and $\min\{f, 1\}$ are in L and $I(\alpha f + \beta g) = \alpha I(f) + \beta I(g)$. To verify (2), let f be a non-negative simple function and $\{f_k\}_{k=1}^{\infty}$ a sequence of non-negative simple functions such that

$$r \equiv I(f) - \sum_k I(f_k) > 0.$$

Let c be a positive upper bound for f, and A an integrable set with $f(x) = 0$ for all $x \in D(f) \cap A^2$. Define the following positive constants:

$$\varepsilon = \min\left\{c, \frac{1}{2} r\ (\mu(A) + 1)^{-1}\right\},$$

$$\alpha = (2c)^{-1}\ (r - 2\varepsilon\ \mu(A)),$$

$$\varepsilon_k = \varepsilon 2^{-k} \qquad (2 \le k < \infty),$$

$$\delta_k = \alpha\varepsilon 2^{-2k-1} \qquad (2 \le k < \infty).$$

By choosing an appropriate sequence $1 = j_1 < j_2 < \cdots$ of integers, and replacing f_k by $\sum \{f_i : j_k \le i < j_{k+1}\}$, we may assume that $I(f_k) \le \delta_k$ for $k \ge 2$. By Lemma 5.6, there exists $A_k \in \mathfrak{M}$ such that $f_k < \varepsilon_k$ on $-A_k$ and

$$\mu(A_k) < 2\varepsilon_k^{-1} I(f_k) \leq \alpha 2^{-k}$$

for $k \geq 2$. Moreover, by Lemma 5.7, there exists $B \in \mathfrak{M}$ such that $f - f_1 > \varepsilon$ on B and

$$\mu(B) \geq c^{-1}\left(I(f - f_1) - 2\varepsilon\,\mu(A)\right)$$
$$\geq c^{-1}\left(r - 2\varepsilon\,\mu(A)\right) > \alpha.$$

Hence

$$\lim_{n\to\infty} \mu(B - A_2 - \ldots - A_n) \geq \mu(B) - \sum_{k=2}^{\infty} \alpha 2^{-k} > \frac{1}{2}\alpha > 0.$$

It follows that $B \wedge \bigwedge_{k=2}^{\infty} - A_k$ contains some element $x \in X$. Then

$$f(x) - f_1(x) - \sum_{k=2}^{\infty} f_k(x) \geq \varepsilon - \sum_{k=2}^{\infty} \varepsilon_k = \frac{1}{2}\varepsilon > 0.$$

To verify (3), choose $M \in \mathfrak{M}$ with $\mu(M) > 0$, and set $p \equiv (\mu(M))^{-1}\chi_M$. Property (4) becomes obvious when a representation $f \equiv \sum_{i=1}^{n} a_i\chi_i$, with $\{\chi_i\}$ the characteristic functions of disjoint sets $\{A_i\}$, is used. ▌

We have seen how a measure space induces an integration space. Now we complete the circle of ideas by showing how that integration space leads back to the measure space. For this purpose, we need the concept of a complete measure space.

<u>Definition 5.9.</u> A measure space $(X, \mathfrak{M}, \mu)$ is <u>complete</u> whenever the following conditions are satisfied:

(1) If $A \in \mathfrak{M}$ and if B is a complemented set with $A^1 \subset B^1$ and $A^2 \subset B^2$, then $B \in \mathfrak{M}$. (We write $A \ \ell e \ B$ whenever $A^1 \subset B^1$ and $A^2 \subset B^2$.)

(2) For each sequence $\{A_k\}$ of integrable sets such that $c \equiv \lim_{n\to\infty} \mu\left(\bigvee_{k=1}^{n} A_k\right)$ exists, the set $\bigvee_k A_k$ belongs to $\mathfrak{M}$ and has measure c.

(3) For complemented sets A, B, C, if A and C are in $\mathfrak{M}$, $\mu(A) = \mu(C)$ and $A < B < C$, then $B \in \mathfrak{M}$ and $\mu(A) = \mu(B) = \mu(C)$. ▌

<u>Lemma 5.10.</u> Let $(X, \mathfrak{M}, \mu)$ be a complete measure space.

(i) If $A \in \mathfrak{M}$ and B is a complemented set such that $A \ \ell e \ B$, then $\mu(B) = \mu(A)$.

(ii) If B is a complemented set and if there exist A and C in $\mathfrak{M}$ such that $x \in B$ whenever $x \in A$, and $x \in -B$ whenever $x \in -C$, and $\mu(A) = \mu(C)$, then $B \in \mathfrak{M}$ and $\mu(A) = \mu(B) = \mu(C)$.

<u>Proof.</u> (i) By (1) of Definition 5.9, B is in $\mathfrak{M}$. Hence

$$\begin{aligned}\mu(B) &= \mu(B \wedge A) + \mu(B - A)\\ &= \mu\left((A^1, A^2)\right) + \mu\left((\phi, A^1 \cup A^2)\right)\\ &= \mu(A).\end{aligned}$$

(ii) Note that $A' \equiv (A^1, A^2 \cup B^2)$ is a complemented set and $A \ \ell e \ A'$. Hence $A' \in \mathfrak{M}$, and $\mu(A') = \mu(A)$. Also, $A' < B$. On the other hand, $C' \equiv (C^1 \cup B^1, C^2)$

is a complemented set and $C \ \ell e \ C'$. Thus $C' \in \mathfrak{m}$, and $\mu(C') = \mu(C)$. Also, $B < C'$. Therefore $B \in \mathfrak{m}$ and

$$\mu(A) = \mu(A') = \mu(B) = \mu(C') = \mu(C)$$

by (3) of Definition 5.9. ▌

There is an equivalent way of stating (3) of Definition 5.9 to make it easier to verify in the proof of Theorem 5.12.

<u>Lemma 5.11.</u> Let $(X, \mathfrak{m}, \mu)$ be a measure space satisfying (1) of Definition 5.9. Then Condition (3) is equivalent to Condition (3′):

(3′) Suppose $N, B \in \mathfrak{m}$ and $\mu(N) = 0$. Let A be a complemented set such that $x \in A$ whenever $x \in B - N$ and $x \in -A$ whenever $x \in -B - N$. Then $A \in \mathfrak{m}$ and $\mu(A) = \mu(B)$.

<u>Proof.</u> Suppose Condition (3) holds and N, B, A are sets as described. Write $C \equiv (B^1 \cap N^2, B^2 \cap N^2)$. Then $B - N < C < B \vee N$. Since $\mu(N) = 0$, we get $\mu(B - N) = \mu(B \vee N) = \mu(B)$. This means $C \in \mathfrak{m}$ and $\mu(C) = \mu(B)$. But $C \ \ell e \ A$. Hence $A \in \mathfrak{m}$ and $\mu(A) = \mu(B)$.

Suppose Condition (3′) holds and $A < B < C$ with $A, C \in \mathfrak{m}$ and $\mu(A) = \mu(C)$. Set $N \equiv C - A$. Then $x \in B$ whenever $x \in A - N$ and $x \in -B$ whenever $x \in -A - N$. Hence $B \in \mathfrak{m}$ and $\mu(B) = \mu(A)$. ▌

<u>Theorem 5.12.</u> Let $(X, \mathfrak{m}, \mu)$ be a measure space and (X, L, I) be the induced integration space. Then $(X, \mathfrak{m}, \mu)$ is complete if and only if every integrable set generated by (X, L, I) is in $\mathfrak{m}$.

<u>Proof.</u> To avoid confusion during the proof, we will stipulate that a complemented set A is called integrable if $\chi_A \in L_1$, that is, there exists a

sequence $\{\varphi_n\}$ in L such that $\sum_n I(|\varphi_n|)$ converges and $\sum_n \varphi_n(x) = \chi_A(x)$ whenever $\sum_n |\varphi_n(x)|$ converges. If $A \in \mathfrak{M}$, then $\chi_A \in L$, so A is integrable, and its measure as a member of $\mathfrak{M}$ is equal to its measure $I(\chi_A)$ as an integrable set.

Suppose $(X, \mathfrak{M}, \mu)$ is complete. Let M be an integrable set whose characteristic function has the representation $\{\varphi_j\}_{j=1}^{\infty}$. Choose a sequence $\{\varepsilon_n\}$ of positive constants with $\sum_n \varepsilon_n < 4^{-1}$. Fix an integer $K \geq 1$. Choose a sequence $\{\delta_n\}$ of positive constants such that

$$2\sum_n \delta_n \varepsilon_n^{-1} < K^{-1}.$$

Choose a sequence

$$1 = j_1 < j_2 < \cdots$$

of integers with $\sum \{I(|\varphi_j|) : j_n \leq j < \infty\} < \delta_n$ for $n \geq 2$. Set $\psi_1 \equiv \sum_{j=1}^{j_2} \varphi_j$. For $n \geq 2$ write

$$\psi_n \equiv \sum \left\{|\varphi_j| : j_n \leq j < j_{n+1}\right\}.$$

Let $\psi_1 \equiv \sum_{i=1}^{m} \alpha_i \chi_i$ be a representation of the simple function ψ_1, where $\{\chi_i\}$ are characteristic functions of disjoint sets $\{C_i\}$. Choose $\alpha \in (\frac{1}{2}, \frac{3}{4})$ with $\alpha \neq \alpha_i$ for each i. Then

$$A_1 \equiv (\{x : \psi_1(x) > \alpha\}, \{x : \psi_1(x) \leq \alpha\})$$

is in $\mathfrak{M}$. By Lemma 5.6, for each $n \geq 2$ there exists $A_n \in \mathfrak{M}$ such that $\psi_n \leq \varepsilon_n$ on $-A_n$ and

$$\mu(A_n) < 2\varepsilon_n^{-1} I(\psi_n) < 2\delta_n \varepsilon_n^{-1}.$$

Hence $B_K \equiv A_1 - \bigvee_{n=2}^{\infty} A_n \in \mathfrak{M}$. For each $x \in B_K$, $x \in D(\chi_M)$, and

$$|\chi_M(x)| = \left|\sum_n \varphi_n(x)\right| \geq \psi_1(x)$$

$$-\sum_{n=2}^{\infty} \psi_n(x) > \alpha - \sum_{n=2}^{\infty} \varepsilon_n > \frac{1}{4},$$

so that $x \in M$. Hence $\mu(M) \geq \mu(B_K)$. Next, assume $\mu\left(M - \bigvee_{n=2}^{\infty} A_n\right) > \mu(A_1)$. Then there exists $y \in M - \bigvee_{n=2}^{\infty} A_n - A_1$. We have $\chi_M(y) = 1$ and

$$\chi_M(y) \leq \sum_n \psi_n(y) < \frac{3}{4} + \sum_{n=2}^{\infty} \psi_n(y) < 1.$$

This is a contradiction. Therefore $\mu\left(M - \bigvee_{n=2}^{\infty} A_n\right) \le \mu(A_1)$. It follows that

$$\mu(M) - 2K^{-1} \le \mu(M) - 2\mu\left(\bigvee_{n=2}^{\infty} A_n\right)$$
$$\le \mu(A_1) - \mu\left(\bigvee_{n=2}^{\infty} A_n\right) \le \mu(B_K).$$

If $\mu(M) < \mu\left(\bigvee_{K=1}^{L} B_K\right)$ for some L, then, there exists y in $\bigvee_{K=1}^{L} B_K - M$, a contradiction. Thus

$$\mu(M) - 2L^{-1} \le \mu\left(\bigvee_{K=1}^{L} B_K\right) \le \mu(M).$$

It follows from (2) of Definition 5.9 that $B \equiv \bigvee_K B_K \in \mathfrak{M}$, and $\mu(B) = \mu(M)$. Also, $x \in M$ for each $x \in B$. In the same way, we set $C_K \equiv A_1 \vee \bigvee_{n=2}^{\infty} A_n$, and prove that $x \in -M$ for each $x \in -C_K$, and that $\mu(M) \le \mu(C_K) \le \mu(M) + 2K^{-1}$. It follows that $C \equiv \bigwedge_K C_K \in \mathfrak{M}$, $\mu(C) = \mu(M)$, and $x \in -M$ for all $x \in -C$. Then Lemma 5.10 gives $M \in \mathfrak{M}$.

Suppose every integrable set is in $\mathfrak{M}$. We have to show that the three conditions of Definition 5.9 are satisfied. To this end, assume $A \in \mathfrak{M}$, and let B be a complemented set with $A \ \ell e \ B$. Then χ_A is an integrable function, and χ_B is an extension of χ_A. Therefore χ_B is also integrable. Hence $B \in \mathfrak{M}$. Thus Condition (1) is satisfied. Proposition 2.9 implies Condition (2). By Lemma 5.11, we may choose to prove Condition (3′) instead of

Condition (3). Hence suppose B and N are in $\mathfrak{M}$, $\mu(N) = 0$, $x \in A$ whenever $x \in B - N$ and $x \in -A$ whenever $x \in -B - N$. Then $\chi_A = \chi_B$ on the full set $(B^1 \cup B^2) \cap N^2$. Therefore A is an integrable set, and $\mu(B) = I(\chi_B) = I(\chi_A) = \mu(A)$. ∎

There is a dual to the above theorem. Starting with an integration space which induces a measure space, this measure space in turn induces an integration space that equals the original one. To state this formally, we need the notion of a complete integration space and a lemma telling us that in a certain sense any integrable function is almost equal to a simple function.

Definition 5.13. An integration space (X, L, I) is complete if $L = L_1$ as subsets of $F(X)$. ∎

Indeed, (X, L_1, I) is a complete integration space by Theorem 1.16, called the completion of (X, L, I).

Lemma 5.14. Let (X, L, I) be any integration space. For each $f \in L_1$ and each $\varepsilon > 0$ there exist integrable sets $A_1, \ldots, A_n$ and constants $c_1, \ldots, c_n$ such that

$$I\left(\left|f - \sum_{i=1}^{n} c_i \chi_i\right|\right) < \varepsilon,$$

where $\{\chi_i\}$ are the characteristic functions of $\{A_i\}$.

Proof. We may take f to be non-negative. Choose an integrable set A and integer N such that $I(f - g) < \frac{1}{2}\varepsilon$, where $g \equiv \min\{\chi_A f, N\}$. Then choose $\delta > 0$ such that $\delta\mu(A) < \frac{1}{2}\varepsilon$, and choose constants $0 = c_0 < c_1 < \cdots < c_n$

such that $c_n > N$, and for $i \in \{1, \ldots, n\}$, $c_i - c_{i-1} < \delta$ and $B_i \equiv (\{x : g(x) \geq c_i\}, \{x : g(x) < c_i\})$ is integrable. Then $A_0 \equiv A - B_1$, $A_1 \equiv B_1 - B_2, \ldots, A_{n-1} \equiv B_{n-1} - B_n$, and $A_n \equiv B_n$ are integrable. Write $h \equiv \sum_{i=0}^n c_i \chi_i$, where $\{\chi_i\}_{i=0}^n$ are the characteristic functions of $\{A_i\}_{i=0}^n$. Then

$$I(|g - h|) = \sum_{i=0}^n I(|g - h| \chi_i)$$

$$\leq \delta \sum_{i=0}^n \mu(A_i) = \delta\mu(A) < \frac{1}{2}\varepsilon.$$

By the triangle inequality, $I(|f - h|) < \varepsilon$. ▌

Theorem 5.15. Let (X, L, I) be an integration space and $(X, \mathfrak{M}, \mu)$ the induced measure space. Then (X, L, I) is complete if and only if every integrable function generated by $(X, \mathfrak{M}, \mu)$ is in L.

Proof. Assume first that (X, L, I) is complete. Let f be an integrable function generated by $(X, \mathfrak{M}, \mu)$. Then f has a representation $\{f_n\}_{n=1}^\infty$, where each f_n is a simple function with respect to $(X, \mathfrak{M}, \mu)$. Hence $f_n \in L_1$ for each n. Since $L = L_1$, $f_n \in L$, so that $\{f_n\}$ is a representation of f by elements of L. Hence $f \in L_1$. Hence $f \in L$.

Assume next that every integrable function generated by $(X, \mathfrak{M}, \mu)$ is in L, and consider $f \in L_1$. By Lemma 5.14, f is an integrable function generated by $(X, \mathfrak{M}, \mu)$. Hence $f \in L$. Thus (X, L, I) is complete. ▌

PRODUCT MEASURES AND FUBINI'S THEOREM

The Daniell technique leads easily to the definition of product measures.

Definition 6.1. Let (X, L_μ, I_μ) and (Y, L_ν, I_ν) be two integration spaces, which respectively induce measures μ on X and ν on Y. Let L be the subset of $F(X \times Y)$ consisting of all functions $f : D(f) \to \mathbb{R}$ of the form

$$f(x, y) \equiv \sum_{k=1}^{n} a_k \chi_k(x) \chi'_k(y),$$

where $a_k \in \mathbb{R}$ and χ_k is the characteristic function of an integrable set $M_k \equiv (M_k^1, M_k^2)$ in X and χ'_k is the characteristic function of an integrable set $N_k \equiv (N_k^1, N_k^2)$ in Y. Clearly $D(f) \equiv \bigcap_{k=1}^{n} (M_k^1 \cup M_k^2) \times (N_k^1 \cup N_k^2)$. The integral $I(f)$ is defined by

$$I(f) \equiv \sum_{k=1}^{n} a_k \mu(M_k) \nu(N_k). \quad \blacksquare$$

To show that $I : L \to \mathbb{R}$ is a function, we need the notion of a simple representation of an element $f \equiv \sum a_k \chi_k \chi'_k$ of L. The representation is called simple if for $j \neq k$, (i) M_k and M_k are either disjoint or equal, (ii) N_j and N_k are either disjoint or equal, (iii) N_j and N_k are disjoint in case $M_j = M_k$ (equivalently, M_j and M_k are disjoint in case $N_j = N_k$). Then every function in L has a simple representation. To see this, consider a function $f = \sum_{k=1}^{n} a_k \chi_k \chi_k'$ in L. For each free subset S of $\{1, \ldots, n\}$, let

$$P(S) \equiv \bigwedge \{M_k : k \in S\} \wedge \bigwedge \{-M_k : k \in -S\}$$

and

$$Q(S) \equiv \bigwedge \{N_k : k \in S\} \wedge \bigwedge \{-N_k : k \in -S\}.$$

For S and T both free subsets of $\{1, \ldots, n\}$, write

$$b_{ST} \equiv \sum \{a_k : k \in S \cap T\}.$$

Then

$$\sum b_{ST} \chi_{P(S)} \chi_{Q(T)}$$

is a simple representation of f. It follows as in Theorem 5.5 that I is a function.

<u>Theorem 6.2.</u> Let (X, L_μ, I_μ) and (Y, L_ν, I_ν) be integration spaces. Define L and I as in Definition 6.1. Then $(X \times Y, L, I)$ is an integration space.

Proof. Using simple representations of functions in L, we easily see that Properties (1), (3) and (4) of Definition 1.1 are satisfied. To verify Property (2), let $\{f_n\}$ be a sequence of non-negative functions in L and f another non-negative function in L such that $\sum_n I(f_n) < I(f)$. Then $I(f_n) = I_\mu(I_\nu(f_n))$ for each n (where $I_\nu(f_n)$ stands for the element $x \to I_\nu(f_n(x, \cdot))$ of $F(X)$), and $I(f) = I_\mu(I_\nu(f))$. Hence $\sum_n I_\mu(I_\nu(f_n)) < I_\mu(I_\nu(f))$. Since (X, L_μ, I_μ) is an integration space, there exists $x_0 \in X$ such that $\sum_n I_\nu(f_n(x_0, \cdot)) < I_\nu(f(x_0, \cdot))$. For the same reason there exists $y_0 \in Y$ such that $\sum_n f_n(x_0, y_0) < f(x_0, y_0)$. ▌

The integration space $(X \times Y, L, I)$ is called the product of (X, L_μ, I_μ) and (Y, L_ν, I_ν). The following basic result on product spaces is known as Fubini's Theorem.

Theorem 6.3. Let f be an integrable function on the product space $(X \times Y, L, I)$. Then for each y belonging to some full subset W of Y, the function $f(\cdot, y)$ is integrable on X. Moreover the function $I_\mu(f)$ with domain W is integrable on Y, with integral $I_\nu(I_\mu(f)) = I(f)$.

Proof. By the hypothesis, there exists a sequence $\{\varphi_n\}$ of functions in L such that

$$f(x, y) = \sum_n \varphi_n(x, y)$$

for all (x, y) in the set $\left\{(x, y) : \sum_n |\varphi_n(x, y)| \text{ converges}\right\}$, for which

$\sum_n I(|\varphi_n|)$ converges. Then $I(f) = \sum_n I(\varphi_n)$. For each n, note that $\zeta_n \equiv I_\mu(|\varphi_n|)$ is a Y-integrable function, and $I_\nu(\zeta_n) = I_\nu(I_\mu(|\varphi_n|)) = I(|\varphi_n|)$. Therefore the series $\sum_n \zeta_n = \sum_n I_\mu(|\varphi_n|)$ converges on some full set $W \in Y$. For each $y_0 \in W$, $f(\cdot, y_0) = \sum_n \varphi_n(\cdot, y_0)$ is an X-integrable function and $I_\mu(f(\cdot, y_0)) = \sum_n I_\mu(\varphi_n(\cdot, y_0))$. Therefore $I_\mu(f)$ exists on the full set W. If we define $\vartheta_n \equiv I_\mu(\varphi_n)$, then ϑ_n is Y-integrable and

$$\sum_n I_\nu(|\vartheta_n|) \leq \sum_n I_\nu(I_\mu(|\varphi_n|)) = \sum_n I(|\varphi_n|).$$

As a result, the series $\sum_n \vartheta_n$ defines a Y-integrable function. But

$$\sum_n \vartheta_n = \sum_n I_\mu(\varphi_n) = I_\mu(f)$$

on the full set $W \cap \bigcap_n D(\vartheta_n)$. Hence $I_\mu(f)$ is Y-integrable. In fact

$$I_\nu(I_\mu(f)) = I_\nu\left(\sum_n \vartheta_n\right) = \sum_n I_\nu(\vartheta_n) = \sum_n I_\nu(I_\mu(\varphi_n)) = \sum_n I(\varphi_n) = I(f). \blacksquare$$

Clearly the above technique can be extended to form the product of any finite number of integration spaces, and to express the integral on the product-space as an iterated integral. These two facts offer a simple way to the construction of the product of a sequence $\{(X_k, L_k, I_k)\}_{k=1}^{\infty}$ of integration spaces, under the condition that $I_k(1) = 1$ for each k, where 1 is the function on X_k which is everywhere equal to 1. More generally, one can consider for each k an integration space (Y_k, L_k, I_k), where $Y_k \equiv \Pi_{i=1}^{k} X_i$, and show that under appropriate consistency conditions it is possible to define an integral on the countable product $\Pi_{n=1}^{\infty} X_n$, inducing the integrals I_k. Since they will not be needed in the sequel, we shall not give these developments here.

APPROXIMATION BY COMPACT SETS IN A LOCALLY COMPACT INTEGRATION SPACE

We have developed an extensive theory of integration spaces without giving any concrete example. It is time for us to present a significant model that contains many special cases.

Definition 7.1. A positive measure μ on a locally compact space X is a linear real-valued function on $C(X)$, the space of all continuous real-valued functions on X with compact support, whose value at an element f of $C(X)$ is written $\int f d\mu$, such that $\int f d\mu \geq 0$ whenever $f \geq 0$. ∎

The elements of $C(X)$ are known as test functions, and to rule out the trivial case, we assume $\int p d\mu \neq 0$ for some $p \in C(X)$.

Our aim now is to show that a positive measure μ on a locally compact space X induces an integration space $(X, C(X), \mu)$. It is easily seen that Properties (1), (3) and (4) of Definition 1.1 are satisfied by $(X, C(X), \mu)$. For instance, to show that

$$\lim_{n\to\infty} I\left(\min \{|f|, n^{-1}\}\right) = 0,$$

define the test function g by $g(x) \equiv (1 - \rho(x, K))^{+}$, where K is the support of f. If we let $M \equiv \sup \{|f(x)| : x \in X\} + 1$, then $0 \leq \min \{|f|, n^{-1}\} \leq n^{-1} Mg$, and clearly $\int n^{-1} Mg \, d\mu \to 0$ as $n \to \infty$. The difficulty lies in showing that Property (2) is also satisfied. To this end, we need two lemmas.

Lemma 7.2. Suppose c, d, δ are real numbers with $c > 0$, $d \geq 0$, $1 > \delta > 0$ and $(1 - 2\delta)c - d > 0$. Then there exists a number

$$\gamma \equiv (4c)^{-1}((1-2\delta)c-d)$$

such that for an arbitrary $m \geq 1$ and arbitrary non-negative numbers $\{c_i\}_{i=1}^m$, $\{d_i\}_{i=1}^m$ with $\sum_{i=1}^m c_i = c$ and $\sum_{i=1}^m d_i = d$, there exists a free subset S of $\{1, \ldots, m\}$ such that $(1-2\delta)c_i - d_i > 0$ for all $i \in S$ and

$$\sum_{i \in S} c_i \geq \gamma \sum_{i=1}^m c_i.$$

<u>Proof.</u> For each $i \in \{1, \ldots, m\}$, either $(1-2\delta)c_i - d_i > 0$ or $(1-2\delta)c_i - d_i < \gamma\, cm^{-1}$. In the former case, let $i \in S$. Otherwise let $i \in T$. Now suppose $\sum_{i \in S} c_i < \gamma \sum_{i=1}^m c_i$. Then

$$\begin{aligned}(1-2\delta)c - d &= \sum_{i \in S} (1-2\delta)c_i - d_i + \sum_{i \in T} (1-2\delta)c_i - d_i \\ &\leq \sum_{i \in S} c_i + \sum_{i \in T} \gamma\, cm^{-1} \\ &< \gamma \sum_{i=1}^m c_i + \gamma c = 2\gamma c \\ &= \frac{1}{2}((1-2\delta)c - d).\end{aligned}$$

This contradiction implies $\sum_{i \in S} c_i \geq \gamma \sum_{i=1}^m c_i$. ∎

Lemma 7.3. Let m be a positive integer, assume that $1 > \delta > 0$, and let $\{b_i, c_i, d_i\}_{i=1}^m$ be non-negative numbers such that

(i) $(1 - 2\delta)c - d > 0$ with $c \equiv \sum_{i=1}^m c_i$ and $d \equiv \sum_{i=1}^m d_i$,

(ii) $b_i \le d_i \quad (1 \le i \le m)$,

(iii) $\sum_{i=1}^m (\delta\gamma c_i - b_i) > 0$ with γ defined as in Lemma 7.2.

Then there exists an integer $k \in \{1, \dots, m\}$ such that $(1 - 2\delta)c_k - d_k > 0$ and $2\delta c_k - b_k > 0$.

Proof. By Lemma 7.2, we choose S such that

$$\sum_{i\in S} c_i \ge \gamma\left(\sum_{i\in S} c_i + \sum_{i\in T} c_i\right)$$

and $(1 - 2\delta)c_i - d_i > 0$ for all $i \in S$. Now assume for each $i \in S$ that $\delta\gamma c_i - b_i < -\delta c_i$. Then

$$0 < \sum_{i=1}^m (\delta\gamma c_i - b_i) = \sum_{i\in S} (\delta\gamma c_i - b_i) + \sum_{i\in T} (\delta\gamma c_i - b_i)$$

$$< \sum_{i\in S} -\delta c_i + \delta\gamma c \le 0.$$

This is a contradiction. Hence for some $k \in S$, $\delta\gamma c_k - b_k \ge -\delta c_k$ or $\delta(1 + \gamma)c_k - b_k \ge 0$. Since $0 < \gamma < 1$ and $c_k > 0$, we have $2\delta c_k - b_k > 0$. ∎

We now complete the demonstration that a positive measure on a locally compact space induces an integration space.

Proposition 7.4. Let μ be a positive measure on a locally compact space X, and g a non-negative test function. Let $\{f_j\}_{j=1}^{\infty}$ be a sequence of non-negative test functions such that $\sum_j \int f_j d\mu$ converges and is less than $\int g d\mu$. Then $\sum_j f_j(x)$ converges for some x in X and $\sum_j f_j(x) < g(x)$.

Proof. We show, by induction on n, that there exist sequences

(a) $\{\lambda_n\}_{n=0}^{\infty}$ of non-negative test functions,

(b) $\{x_n\}_{n=1}^{\infty}$ of points in X,

(c) $\{P(n)\}_{n=0}^{\infty}$ of integers such that $1 \equiv P(0) < P(1) < P(2) < \dots$,

(d) $\{\delta_n\}_{n=0}^{\infty}$ of positive real numbers with $\delta_0 \equiv 2^{-1}$ and $\delta_n \le 2^{-n}$ for $n \ge 1$,

(e) $\{\theta_n\}_{n=1}^{\infty}$ of positive real numbers with $\theta_n < n^{-2}$,

such that, for some $\varepsilon > 0$ and all positive integers m and n, we have

(1) $\rho(x_m, x_n) \le \theta_m + \theta_n$, where ρ is the metric on X,

(2) $\lambda_n(x_n) > 0$, and $\lambda_n(y) = 0$ for all y with $\rho(x_n, y) \ge \theta_n$,

(3) $(1 - 2\delta_n)\left(\prod_{k=0}^{n-1} 2\delta_k\right) \int (g - \varepsilon)\lambda_n d\mu - \sum_{j=P(n-1)}^{P(n)-1} \int f_j \lambda_n d\mu > 0$,

(4) $\left(\prod_{k=0}^{n} 2\delta_k\right) \int (g - \varepsilon)\lambda_n d\mu - \sum_{j=P(n)}^{\infty} \int f_j \lambda_n d\mu > 0$,

(5) $\theta_n < \frac{1}{3}\omega_n\left(\frac{1}{2}\varepsilon(1 - 2\delta_n)\left(\prod_{k=0}^{n-1} 2\delta_k\right)\right)$,

where ω_n is any common modulus of continuity of the functions g and $f_{P(n-1)} + \cdots + f_{P(n)-1}$, with $\omega_n \leq \omega_{n-1}$ for all $n \geq 1$.

Let λ_0 be any non-negative test function, with $0 \leq \lambda_0 \leq 1$, such that $\lambda_0 g = g$. By hypothesis, there exists $\varepsilon > 0$ such that

$$\int (g - \varepsilon)\, \lambda_0 d\mu - \sum_j \int f_j\, \lambda_0 d\mu > 0,$$

so that (4) holds for $n = 0$.

Now let n be any positive integer. Suppose that for all $k < n$, the elements λ_k, x_k, $P(k)$, δ_k and θ_k have been constructed. Let $c \equiv \left(\prod_{k=0}^{n-1} 2\delta_k\right) \int (g - \varepsilon)\, \lambda_{n-1} d\mu$ and $d \equiv \sum_{j=P(n-1)}^{\infty} \int f_j\, \lambda_{n-1} d\mu$. Then $c - d > 0$ by (4). Hence there exists a positive number $\delta_n \leq 2^{-n}$ such that

$$(1 - 2\delta_n)c - d > 0. \tag{6}$$

Let $\gamma_n \equiv (4c)^{-1}\, ((1 - 2\delta_n)c - d)$. Choose $P(n) > P(n - 1)$ such that

$$\sum_{j=P(n)}^{\infty} \int f_j\, \lambda_{n-1} d\mu < \delta_n \gamma_n c.$$

Let K be the support of λ_{n-1}. Note that $\operatorname{diam}(K) > 0$. Let

$$\theta_n \equiv \min \left\{ \frac{1}{2} n^{-2}, \frac{1}{3} \omega_n \left(\frac{1}{2} \varepsilon \ (1 - 2\delta_n) \left(\prod_{k=0}^{n-1} 2\delta_k \right) \right), \text{ diam } (K) \right\}$$

and $\{y_1, \ldots, y_m\}$ be a $\frac{1}{4}\theta_n$ approximation to K. Choose a constant α in $\left(\frac{1}{4}\theta_n, \frac{1}{2}\theta_n\right)$ so that the sets

$$K_i \equiv \{ y \in K : \rho(y, y_i) \geq \alpha \} \qquad (1 \leq i \leq m)$$

are compact. (See Theorem 8 of Chapter 4 of [2], or Theorem 7.7 below.) For each i $(1 \leq i \leq m)$ let λ^i be the test function defined by

$$\lambda^i(y) \equiv \lambda_{n-1}(y) \ \rho(y, K_i) \left[\sum_{j=1}^{m} \rho(y, K_j) \right]^{-1}.$$

Then $0 \leq \lambda^i \leq \lambda_{n-1}$, $\sum_{i=1}^m \lambda^i = \lambda_{n-1}$, and $\lambda^i(y) = 0$ whenever $\rho(y, y_i) \geq \alpha$. By (6) we have

$$\sum_{i=1}^{m} \left\{ (1 - 2\delta_n) \prod_{k=0}^{n-1} 2\delta_k \int (g - \varepsilon)\lambda^i d\mu - \sum_{j=P(n-1)}^{P(n)-1} \int f_j \lambda^i d\mu \right.$$

$$\left. - \sum_{j=P(n)}^{\infty} \int f_j \lambda^i d\mu \right\} > 0.$$

By Lemma 7.3, there exists i in $\{1, \ldots, m\}$ with

$$(1 - 2\delta_n) \prod_{k=0}^{n-1} 2\delta_k \int (g - \varepsilon) \lambda^i d\mu - \sum_{j=P(n-1)}^{P(n)-1} \int f_j \lambda^i d\mu > 0$$

and

$$2\delta_n \prod_{k=0}^{n-1} 2\delta_k \int (g - \varepsilon) \lambda^i d\mu - \sum_{j=P(n)}^{\infty} \int f_j \lambda^i d\mu > 0.$$

Define $\lambda_n \equiv \lambda^i$, and choose x_n with $\lambda_n(x_n) > 0$. With patience, one finds that conditions (2) through (5) are satisfied, and condition (1) follows from condition (2). This completes the construction.

By (1), we see that $\{x_n\}$ is a Cauchy sequence whose limit x satisfies the inequalities $\rho(x, x_n) \le \theta_n$. By (2), we see that $\lambda_n(y) = 0$ whenever $\rho(y, x) > 2\theta_n$.

Assume that

$$\sum_{j=P(n-1)}^{P(n)-1} f_j(x) > (1 - 2\delta_n) \prod_{k=0}^{n-1} 2\delta_k \left(g(x) - \tfrac{1}{4}\varepsilon\right).$$

Then (5) implies that for all points y with $\rho(y, x) \le 3\theta_n$,

$$\sum_{j=P(n-1)}^{P(n)-1} f_j(y) \ge (1 - 2\delta_n) \prod_{k=0}^{n-1} 2\delta_k \left(g(y) - \varepsilon\right).$$

Since $\lambda_n(y) = 0$ whenever $\rho(y, x) > 2\theta_n$, we have

$$\sum_{j=P(n-1)}^{P(n)-1} f_j \lambda_n \geq (1 - 2\delta_n) \prod_{k=0}^{n-1} 2\delta_k (g - \varepsilon)\lambda_n$$

and therefore

$$\sum_{j=P(n-1)}^{P(n)-1} \int f_j \lambda_n d\mu \geq (1 - 2\delta_n) \prod_{k=0}^{n-1} 2\delta_k \int (g - \varepsilon) \lambda_n d\mu.$$

This inequality contradicts (3). Hence

$$\sum_{j=P(n-1)}^{P(n)-1} f_j(x) \leq (1 - 2\delta_n) \prod_{k=0}^{n-1} 2\delta_k \left(g(x) - \frac{1}{4} \varepsilon\right).$$

Since $\sum_n \left\{(1 - 2\delta_n) \prod_{k=0}^{n-1} 2\delta_k\right\}$ converges and is less than 1, we conclude that $\sum_j f_j(x)$ converges and is less than $g(x)$. ▌

The rest of this section will be devoted to proving a stronger version of the theorem that any integrable set of positive measure can be approximated from within by a compact integrable set [2]. A complemented set A is compact if A^1 is compact as a subset of X, and A is closed if A^1 is closed.

Definition 7.5. A compact set K in a locally compact space X is said to be strongly integrable with respect to a positive measure μ on X if there exists a constant c such that for each $\varepsilon > 0$, there exists $\delta > 0$ such that

$$\left| \int f d\mu - c \right| \le \varepsilon$$

for every test function f with $0 \le f \le 1$, $f(x) = 1$ if $x \in K$ and $f(x) = 0$ if $\rho(x, K) \ge \delta$. ▌

The constant c is uniquely defined. In fact, K being strongly integrable implies that the complemented set $(K, -K)$ is integrable with respect to μ, and $\mu((K, -K)) = c$. The proofs are left to the reader.

The next step is to use the theory of profiles to show that there are many strongly integrable sets.

Lemma 7.6. Let $h : K \to \mathbb{R}$ be a continuous function on a compact space K. Let $x_0 \in K$. Suppose $b \equiv \sup \{h(x) : x \in K\} > a \equiv \inf \{h(x) : x \in K\}$. Then for all except countably many real numbers t in (a, b) the distance $\rho(x_0, K^t)$ exists, where $K^t \equiv \{x \in K : h(x) \ge t\}$. Moreover, if t is admissible (is not any of the countably many exceptional values) and ε is any positive constant, then there exists $\delta > 0$ such that

$$|\rho(x_0, K^\alpha) - \rho(x_0, K^t)| < \varepsilon$$

whenever α is admissible and $|\alpha - t| < \delta$.

Proof. Let x_1 be any point in K with $h(x_1) = c$. Let F denote the set of all continuous functions $f : [a, b] \to [0, 1]$. For each $f \in F$, define $\lambda(f) \equiv - \inf \{\max \{\rho(x, x_0), \rho(x_1, x_0)(1 - f(h(x)))\} : x \in K\}$. Then (F, λ) is a profile over $[a, b]$. Let t be a point in (a, c) where the profile is smooth. Therefore there exists a number $\overline{\lambda}(t)$ such that for each $\zeta > 0$, there exists $\eta > 0$ such that $|\lambda(f) - \overline{\lambda}(t)| < \zeta$ for each f in F that satisfies the conditions (a) $f(u) = 0$ if $u \le t - \eta$ and (b) $f(u) = 1$ if $u \ge t + \eta$. Choose f_1 in F so that $f_1(u) = 0$ if $u \le t + \frac{1}{2}\eta$ and $f_1(u) = 1$ if $u \ge t + \eta$. For an arbitrary point $x \in K$, either $h(x) < t + \frac{1}{2}\eta$ or $h(x) > t$. In the former case,

$$\max \{\rho(x, x_0), \rho(x_1, x_0)(1 - f_1(h(x)))\} \ge \rho(x_1, x_0).$$

Since $\max \{\rho(x_1, x_0), \rho(x_1, x_0)(1 - f_1(h(x_1)))\} = \rho(x_1, x_0)$, we get

$$-\lambda(f_1) = \inf \{\max \{\rho(x, x_0), \rho(x_1, x_0)(1 - f_1(h(x)))\} : x \in K^t\}.$$

Since $|\lambda(f_1) - \overline{\lambda}(t)| < \zeta$, there exists $z \in K^t$ such that

$$\rho(z, x_0) \le \max \{\rho(z, x_0), \rho(x_1, x_0)(1 - f_1(h(z)))\} < -\overline{\lambda}(t) + \zeta.$$

On the other hand, choose f_2 in F with $f_2(u) = 0$ if $u \le t - \eta$ and $f_2(u) = 1$ if $u \ge t - \frac{1}{2}\eta$. Then for any $z \in K^t$,

$$\begin{aligned}\rho(z, x_0) &= \max \{\rho(z, x_0), \rho(x_1, x_0)(1 - f_2(h(z)))\} \\ &\ge -\lambda(f_2) > -\overline{\lambda}(t) - \zeta.\end{aligned}$$

Since ζ is arbitrary, we conclude that $\rho(x_0, K^t)$ exists and equals $-\overline{\lambda}(t)$.

If ε is any positive constant, choose $\theta > 0$ such that $|\lambda(f) - \overline{\lambda}(t)| < \frac{1}{2}\varepsilon$ for each $f \in F$ such that (a) $f(u) = 0$ whenever $u \leq t - \theta$ and (b) $f(u) = 1$ whenever $u \geq t + \theta$. Let $\delta \equiv \frac{1}{2}\theta$ and consider any admissible α with $|\alpha - t| < \delta$. Then we can find $\gamma < \delta$ such that $|\lambda(g) - \overline{\lambda}(\alpha)| < \frac{1}{2}\varepsilon$ for each $g \in F$ satisfying (a) $g(x) = 0$ whenever $x \leq \alpha - \gamma$ and (b) $g(x) = 1$ whenever $x \geq \alpha + \gamma$. Therefore

$$|\rho(x_0, K^{\alpha}) - \rho(x_0, K^t)| = |\overline{\lambda}(\alpha) - \overline{\lambda}(t)|$$

$$\leq |\lambda(g) - \lambda(\alpha)| + |\lambda(g) - \overline{\lambda}(t)| < \varepsilon. \quad \blacksquare$$

Theorem 7.7. Let $h : K \to \mathbb{R}$ be a continuous function on a compact space K and suppose

$$b \equiv \sup\{h(x) : x \in K\} > a \equiv (\inf\{h(x) : x \in K\})^+.$$

Then for all except countably many real numbers β in (a, b) the set

$$K^{\beta} \equiv \{x \in K : h(x) \geq \beta\}$$

is compact.

Proof. Let $\{x_n\}_{n=1}^{\infty}$ be a countable dense subset of K. For each n, $\rho(x_n, K^t)$ exists for all except countably many t in (a, b). It follows that $\rho(x_n, K^t)$ exists for all except countably many t in (a, b) and for all $n \in \mathbb{Z}^+$. Therefore K^t is located for each admissible t. Since it is clearly closed, it is compact. ▌

Theorem 7.8. Let $g : X \to \mathbb{R}$ be a test function on a locally compact space X and suppose

$$b \equiv \sup \{g(x) : x \in X\} > 0.$$

Then for all except countably many real numbers β in $(0, b)$ the set

$$X^\beta \equiv \{x \in X : g(x) \geq \beta\}$$

is strongly integrable.

Proof. Let K be the compact support of g. Then the restriction h of g to K satisfies the hypotheses of Theorem 7.7. Therefore for all except countably many β in $(0, b)$ the set

$$K^\beta \equiv \{x \in K : h(x) = g(x) \geq \beta\}$$

is compact. Therefore X^β is compact because $X^\beta = K^\beta$. By Theorem 3.5, X^β is strongly integrable for all except countably many such β. ▌

In the following, we often let a compact set A stand for the complemented set $(A, -A)$, without further discussion.

Definition 7.9. Let ε be a positive constant. An integrable set B is *ε-strongly integrable* if there exists a strongly integrable set L such that $B^1 \subset L^1$ and $\mu(L) - \mu(B) < \varepsilon$. ▌

Lemma 7.10. Let A be any integrable set of positive measure. Let ε be a positive constant. Then there exists a closed ε-strongly integrable set C, a finite set H, and a positive constand $d < \varepsilon$, such that $\mu(A) - \mu(C) < \varepsilon$, $C^1 \subset A^1$, $\mu(Sc(y, d) \cap C) > 0$ for all y in H, and $\rho(x, H) < \varepsilon$ for all x in C.

Proof. We may assume that $\varepsilon < \frac{1}{2}$. Let χ be the characteristic function of A, and let $\{\varphi_n\}$ be a sequence of test functions that represents χ. Then there exists a sequence $1 < N(1) < N(2) < \ldots$ of integers such that

$$\text{(a)} \qquad \int \left|\chi - \sum_{k=1}^{N(1)-1} \varphi_k\right| d\mu < \frac{1}{8}\varepsilon$$

and

$$\text{(b)} \qquad \sum_{k=N(j)}^{\infty} \int |\varphi_n| \, d\mu < \varepsilon^2 \, 2^{-2j-6} \qquad (j \geq 1).$$

Now consider the test functions

$$g_1 \equiv \sum_{k=1}^{N(1)-1} \varphi_k$$

and

$$g_j = \sum_{k=N(j-1)}^{N(j)-1} |\varphi_k| \qquad (j \geq 2).$$

Choose c in $(\frac{1}{2}\varepsilon, \varepsilon)$ such that

$$B_1 \equiv \{x : g_1(x) \geq c\}$$

is strongly integrable, and $-B_j$ $(j \geq 2)$ is integrable, where

$$B_j \equiv (\{x : g_j(x) \leq 2^{-j-2} c\}, \{x : g_j(x) > 2^{-j-2} c\}).$$

Then for $j \geq 2$,

$$2^{-j-2} c\, \mu(-B_j) \leq \int g_j d\mu < \varepsilon^2\, 2^{-2j-4}$$

and hence

$$\mu(-B_j) < \varepsilon 2^{-j-1}.$$

Therefore $B \equiv B_1 - \bigvee_{j=2}^{\infty} - B_j = \bigwedge_j B_j$ is integrable and

$$\mu(B) \geq \mu(B_1) - \mu\left(\bigvee_{j=2}^{\infty} - B_j\right) \geq \mu(B_1) - \sum_{j=2}^{\infty} \mu(-B_j) \geq \mu(B_1) - \frac{1}{4}\varepsilon.$$

Moreover,

$$(1 - \varepsilon)(\mu(A) - \mu(B_1)) \le (1 - \varepsilon)(\mu(A - B_1)) \le \int |\chi - g_1| \, d\mu < \frac{1}{8}\varepsilon.$$

Hence $\mu(A) - \mu(B_1) < \frac{1}{8}(1 - \varepsilon)^{-1}\varepsilon < \frac{1}{4}\varepsilon$ and thus $\mu(A) - \mu(B) < \frac{1}{2}\varepsilon$.

To show that $B^1 \subset A^1$, let $x \in B^1$. Then $\sum_k |\varphi_k(x)|$ converges and

$$\chi(x) = \sum_k \varphi_k(x) \ge g_1(x) - \sum_{k=2}^{\infty} g_j(x) \ge c - \frac{1}{2}\varepsilon > 0.$$

Hence, $x \in A$.

B^1 is closed because B_j^1 is closed for each j. Let the finite subset F of B_1 be a $\frac{1}{4}\varepsilon$-approximation to B_1. Choose d in $(\frac{1}{4}\varepsilon, \frac{1}{2}\varepsilon)$ such that $Sc(y, d) \equiv \{x : \rho(x, y) \le d\}$ is strongly integrable for all y in F. Then $B^1 \subset B_1^1 \subset \bigvee\{Sc(y, d) : y \in F\}$. Write F as a union of disjoint sets H and H', each of which is either finite or void, with $\mu(B \wedge Sc(y, d)) > 0$ for all y in H and $\mu(B \wedge Sc(y, d)) < \frac{1}{2}M^{-1}\varepsilon$ for all y in H', where M is the number of elements in F. Choose α in $(2d, \varepsilon)$ such that

$$S \equiv \{x \in B_1 : \rho(x, H) \le \alpha\} = \{x \in B_1 : (1 - \rho(x, H))^+ \ge 1 - \alpha\}$$

is strongly integrable. The set

$$C \equiv B \wedge S$$

is integrable, C^1 is closed, $C^1 \subset B^1 \subset A^1$, and

$$\mu(B) - \mu(C) \leq \mu\Big(B \wedge \big[\bigvee \{Sc(y, d) : y \in F\} - \bigvee \{Sc(y, d) : y \in H\}\big]\Big)$$

$$\leq \sum \{\mu(B \wedge Sc(y, d) : y \in H'\} < M(\tfrac{1}{2} M^{-1} \varepsilon) = \tfrac{1}{2} \varepsilon.$$

Hence $\mu(A) - \mu(C) = \mu(A) - \mu(B) + \mu(B) - \mu(C) < \varepsilon$. If $x \in C$, then $x \in S$ and $\rho(x, H) < \alpha < \varepsilon$. If $y \in H$ then $\mu(Sc(y, d) \cap C) = \mu(Sc(y, d) \cap B) > 0$. Finally C is ε-strongly integrable because $C^1 \subset S^1$, S is strongly integrable, and

$$\mu(S) - \mu(C) = \mu(S) - \mu(B) + \mu(B) - \mu(C)$$

$$\leq \mu(B_1) - \mu(B) + \mu(B) - \mu(C) < \varepsilon. \quad \blacksquare$$

<u>Theorem 7.11.</u> Let A be an integrable set of positive measure relative to a positive measure μ on a locally compact space X. Then there exists a strongly integrable set $K < A$ such that $\mu(A) - \mu(K)$ is arbitrarily small.

<u>Proof.</u> Let $\varepsilon < 1$ be any positive constant. By Lemma 7.10, we construct a sequence $\{C_k\}$ of integrable sets with

$$A^1 \supset C_1^1 \supset C_2^1 \supset \ldots ,$$

a sequence $\{d_k\}$ of positive constants converging to 0, a sequence $\{\varepsilon_k\}$ of positive constants, and a sequence $\{H_k\}$ of finite sets such that C_k is closed and $\varepsilon 2^{-k}$ strongly-integrable, with

$$\mu(C_k) - \mu(C_{k+1}) < \varepsilon\, 2^{-(k+1)},$$

such that $\rho(x, H_k) < \varepsilon\, 2^{-k}$ for all x in C_k, and such that $\mu(Sc(y, d_k) \cap C_j) > \varepsilon_k$ for all y in H_k and all $j \geq k$. In addition, we choose C_1 with $\mu(A) - \mu(C_1) < \frac{1}{2}\varepsilon$. Write $J \equiv \bigwedge_k C_k$. It follows that $\mu(Sc(y, d_k) \cap J) \geq \varepsilon_k$ for each k and each y in H_k. Thus there exists a point y' in J with $\rho(y, y') \leq d_k$. Since also $\rho(x, H_k) < \varepsilon\, 2^{-k}$ for all x in J, it follows that $\{y' : y \in H_k\}$ is a $d_k + \varepsilon\, 2^{-k}$ approximation to J^1. Therefore J^1 is totally bounded. It is closed because each C_k^1 is closed. It is therefore compact. To show that J^1 is strongly integrable, let $\alpha > 0$ be given. Choose an integer $m > 1$ such that $\varepsilon\, 2^{-m} < \frac{1}{3}\alpha$ and $\mu(C_m) - \mu(J) < \frac{1}{3}\alpha$. Let L_m be a strongly integrable set such that

$$C_m^1 \subset L_m^1 \text{ and } \mu(L_m) - \mu(C_m) < \varepsilon\, 2^{-m}.$$

Choose $\beta > 0$ such that

$$0 \leq \int f d\mu - \mu(L_m) < \frac{1}{3}\alpha$$

for all $f \in C(X)$ satisfying the conditions: $0 \leq f \leq 1$, $f(x) = 1$ if $x \in L_m$ and $f(x) = 0$ if $\rho(x, L_m) \geq \beta$. Since $J^1 \subset L_m^1$, for every test function g with $0 \leq g \leq 1$, $g(x) = 1$ if $x \in J$ and $g(x) = 0$ if $\rho(x, J) \geq \beta$, we have

$$0 \leq \int g d\mu - \mu(J) \leq \int g d\mu - \mu(L_m) + \mu(L_m) - \mu(C_m) + \mu(C_m) - \mu(J)$$

$$< \frac{1}{3}\alpha + \frac{1}{3}\alpha + \frac{1}{3}\alpha = \alpha.$$

Therefore the complemented set $K \equiv (J^1, -J^1)$ is strongly integrable and $\mu(K) = \mu(J)$. By Lemma 6 of Chapter 6 of [2], $K < A$. Finally we compute

$$\mu(K) \geq \mu(A) - (\mu(A) - \mu(C_1)) - \sum_k (\mu(C_k) - \mu(C_{k+1}))$$

$$> \mu(A) - \sum_k \varepsilon 2^{-k} = \mu(A) - \varepsilon. \quad \blacksquare$$

REFERENCES

[1] Asplund, Edgar and Bungart, Lutz, *A First Course in Integration*, Holt, Rinehart and Winston, New York, 1966.

[2] Bishop, Errett, *Foundations of Constructive Analysis*, McGraw-Hill Book Company, New York, 1967.

[3] Chan, Yuen-Kwok, "Aspects of Constructive Measure and Probability Theory", Thesis, University of California at San Diego, 1969.

[4] Daniell, P. J., "A General Form of Integral", *Ann. of Math.* (2) 19(1919), 279-294.